Patricia Melin, Janusz Kacprzyk, and Witold Pedrycz (Eds.)

Bio-inspired Hybrid Intelligent Systems for Image Analysis and Pattern Recognition

Studies in Computational Intelligence, Volume 256

Editor-in-Chief
Prof. Janusz Kacprzyk
Systems Research Institute
Polish Academy of Sciences
ul. Newelska 6
01-447 Warsaw
Poland
E-mail: kacprzyk@ibspan.waw.pl

Further volumes of this series can be found on our homepage: springer.com

Vol. 235. Reiner Onken and Axel Schulte
System-Ergonomic Design of Cognitive Automation, 2009
ISBN 978-3-642-03134-2

Vol. 236. Natalio Krasnogor, Belén Melián-Batista, José A. Moreno-Pérez, J. Marcos Moreno-Vega, and David Pelta (Eds.)
Nature Inspired Cooperative Strategies for Optimization (NICSO 2008), 2009
ISBN 978-3-642-03210-3

Vol. 237. George A. Papadopoulos and Costin Badica (Eds.)
Intelligent Distributed Computing III, 2009
ISBN 978-3-642-03213-4

Vol. 238. Li Niu, Jie Lu, and Guangquan Zhang
Cognition-Driven Decision Support for Business Intelligence, 2009
ISBN 978-3-642-03207-3

Vol. 239. Zong Woo Geem (Ed.)
Harmony Search Algorithms for Structural Design Optimization, 2009
ISBN 978-3-642-03449-7

Vol. 240. Dimitri Plemenos and Georgios Miaoulis (Eds.)
Intelligent Computer Graphics 2009, 2009
ISBN 978-3-642-03451-0

Vol. 241. János Fodor and Janusz Kacprzyk (Eds.)
Aspects of Soft Computing, Intelligent Robotics and Control, 2009
ISBN 978-3-642-03632-3

Vol. 242. Carlos Artemio Coello Coello, Satchidananda Dehuri, and Susmita Ghosh (Eds.)
Swarm Intelligence for Multi-objective Problems in Data Mining, 2009
ISBN 978-3-642-03624-8

Vol. 243. Imre J. Rudas, János Fodor, and Janusz Kacprzyk (Eds.)
Towards Intelligent Engineering and Information Technology, 2009
ISBN 978-3-642-03736-8

Vol. 244. Ngoc Thanh Nguyen, Rados law Piotr Katarzyniak, and Adam Janiak (Eds.)
New Challenges in Computational Collective Intelligence, 2009
ISBN 978-3-642-03957-7

Vol. 245. Oleg Okun and Giorgio Valentini (Eds.)
Applications of Supervised and Unsupervised Ensemble Methods, 2009
ISBN 978-3-642-03998-0

Vol. 246. Thanasis Daradoumis, Santi Caballé, Joan Manuel Marquès, and Fatos Xhafa (Eds.)
Intelligent Collaborative e-Learning Systems and Applications, 2009
ISBN 978-3-642-04000-9

Vol. 247. Monica Bianchini, Marco Maggini, Franco Scarselli, and Lakhmi C. Jain (Eds.)
Innovations in Neural Information Paradigms and Applications, 2009
ISBN 978-3-642-04002-3

Vol. 248. Chee Peng Lim, Lakhmi C. Jain, and Satchidananda Dehuri (Eds.)
Innovations in Swarm Intelligence, 2009
ISBN 978-3-642-04224-9

Vol. 249. Wesam Ashour Barbakh, Ying Wu, and Colin Fyfe
Non-Standard Parameter Adaptation for Exploratory Data Analysis, 2009
ISBN 978-3-642-04004-7

Vol. 250. Raymond Chiong and Sandeep Dhakal (Eds.)
Natural Intelligence for Scheduling, Planning and Packing Problems, 2009
ISBN 978-3-642-04038-2

Vol. 251. Zbigniew W. Ras and William Ribarsky (Eds.)
Advances in Information and Intelligent Systems, 2009
ISBN 978-3-642-04140-2

Vol. 252. Ngoc Thanh Nguyen and Edward Szczerbicki (Eds.)
Intelligent Systems for Knowledge Management, 2009
ISBN 978-3-642-04169-3

Vol. 253. Akitoshi Hanazawa, Tsutom Miki, and Keiichi Horio (Eds.)
Brain-Inspired Information Technology, 2009
ISBN 978-3-642-04024-5

Vol. 254. Kyandoghere Kyamakya, Wolfgang A. Halang, Herwig Unger, Jean Chamberlain Chedjou, Nikolai F. Rulkov, and Zhong Li (Eds.)
Recent Advances in Nonlinear Dynamics and Synchronization, 2009
ISBN 978-3-642-04226-3

Vol. 255. Catarina Silva and Bernardete Ribeiro
Inductive Inference for Large Scale Text Classification, 2009
ISBN 978-3-642-04532-5

Vol. 256. Patricia Melin, Janusz Kacprzyk, and Witold Pedrycz (Eds.)
Bio-inspired Hybrid Intelligent Systems for Image Analysis and Pattern Recognition, 2009
ISBN 978-3-642-04515-8

Patricia Melin, Janusz Kacprzyk,
and Witold Pedrycz (Eds.)

Bio-inspired Hybrid Intelligent Systems for Image Analysis and Pattern Recognition

Springer

Prof. Patricia Melin
Tijuana Institute of Technology
Department of Computer Science
P.O. Box 4207
Chula Vista CA 91909
USA
E-mail: epmelin@hafsamx.org

Prof. Witold Pedrycz
University of Alberta
Dept. Electrical and Computer Engineering
Edmonton, Alberta T6J 2V4
Canada
E-mail: pedrycz@ee.ualberta.ca

Prof. Janusz Kacprzyk
Polish Academy of Sciences
Systems Research Institute
Newelska 601-447 Warszawa
Poland
E-mail: kacprzyk@ibspan.waw.pl

ISBN 978-3-642-26065-0 e-ISBN 978-3-642-04516-5

DOI 10.1007/978-3-642-04516-5

Studies in Computational Intelligence ISSN 1860-949X

Softcover reprint of the hardcover 1st edition 2010

Typeset & Cover Design: Scientific Publishing Services Pvt. Ltd., Chennai, India.

Printed in acid-free paper

9 8 7 6 5 4 3 2 1

springer.com

Preface

We describe in this book, new methods bio-inspired models and applications of hybrid intelligent systems using soft computing techniques for image analysis and pattern recognition. Soft Computing (SC) consists of several intelligent computing paradigms, including fuzzy logic, neural networks, and evolutionary algorithms, which can be used to produce powerful hybrid intelligent systems. The book is organized in four main parts, which contain a group of papers around a similar subject. The first part consists of papers with the main theme of introductory concepts and models, which are basically papers that propose new fuzzy and bio-inspired models to solve general problems. The second part contains papers with the main theme of modular neural networks in pattern recognition, which are basically papers using bio-inspired techniques, like modular neural networks, for achieving pattern recognition in different applications. The third part contains papers with the themes of modular neural networks and other soft computing models in time series prediction, which are papers that apply hybrid intelligent systems to the problem of time series analysis and prediction. The fourth part contains papers that deal with bio-inspired models in optimization and robotics applications.

In the part of Introductory Concepts and Models there are 3 papers that describe different contributions on fuzzy logic and bio-inspired models with application in image analysis and pattern recognition. The first paper, by Humberto Bustince et al., deals with a survey of applications of the extensions of fuzzy sets to image processing. The second paper, by Cecilia Leal-Ramirez et al., deals with an interval type-2 fuzzy cellular model applied to the dynamics of a uni-specific population induced by environment variations. The third paper, by Gustavo Olague and Leonardo Trujillo, studies a genetic programming approach to the design of interest point operators in images.

In the part of Modular Neural Networks in Pattern Recognition there are 5 papers that describe different contributions on achieving pattern recognition using hybrid intelligent systems. The first paper, by Ricardo Munoz et al., describes modular neural networks with fuzzy logic integration for face, fingerprint and voice recognition. The second paper, by Monica Beltran et al., deals with modular neural networks with fuzzy response integration for human signature recognition. The third paper, by Magdalena Serrano et al., proposes an intelligent hybrid system for person identification using biometric measures and modular neural networks with fuzzy integration of responses. The fourth paper, by Denisse Hidalgo et al., describes the optimization of modular neural networks with type-2 fuzzy integration using a general evolutionary method with application in multimodal biometry. The fifth paper, by Miguel Lopez et al., deals with a comparative study of fuzzy methods for response integration in ensemble neural networks for pattern recognition.

In the part of Neural Network Models and Time Series Prediction there are 4 papers that describe different contributions of new neural network models and their application in solving time series prediction problems. The first paper by Martha Pulido et al., describes a hybrid model based on ensemble neural networks and fuzzy logic for aggregation of results for time series prediction in the Mackey-Glass problem. The second paper, by Gerardo Mendez and Angeles Hernandez, deals with the prediction of the MXNUSD exchange rate using a hybrid interval type-2 fuzzy logic system as a forecasting tool. The third paper, by Eduardo Gomez-Ramirez et al., describes a process for discovering universal polynomial cellular neural networks through genetic algorithms. The fourth paper by Mario Chacon et al., describes an EMG hand burst activity detection study based on hard and soft thresholding.

In the part of Optimization and Robotics several contributions are described on the application of bio-inspired methods for achieving optimization and robotics applications. The first paper, by Fevrier Valdez et al., describes a new evolutionary method combining particle swarm optimization and genetic algorithms using fuzzy logic, and its application in optimizing the architecture of modular neural networks. The second paper, by Cynthia Solano-Aragon and Arnulfo Alanis, describes a new multi-agent architecture for controlling autonomous mobile robots using fuzzy logic and obstacle avoidance with computer vision techniques. The third paper, by Ricardo Martinez-Marroquin et al., deals with the Optimization of type-1 and type-2 fuzzy logic controllers of autonomous mobile robots using particle swarm optimization.

In conclusion, the edited book comprises papers on diverse aspects of bio-inspired models, soft computing and hybrid intelligent systems. There are theoretical aspects as well as application papers.

June 30, 2009

Patricia Melin
Tijuana Institute of Technology, Mexico

Janusz Kacprzyk
Polish Academy of Sciences, Poland

Witold Pedrycz
University of Alberta, Canada

Contents

Part I
Introductory Concepts and Models

A Survey of Applications of the Extensions of Fuzzy Sets to Image Processing

Humberto Bustince[1], Miguel Pagola[1], Aranzazu Jurio[1], Edurne Barrenechea[1], Javier Fernández[1], Pedro Couto[2], and Pedro Melo-Pinto[2]

[1] Universidad Pública de Navarra, Pamplona, Spain
miguel.pagola@unavarra.es
[2] Universidade Tras-os-Montes e Alto Douro, VilaReal, Portugal
pmelo@utad.pt

Abstract. In this chapter a revision of different image processing applications developed with different extensions of fuzzy sets is presented. The way extensions of fuzzy sets try to modelize some aspects of the uncertainty existing in different processes of image processing and how this extensions handle in a better way than fuzzy sets such uncertainty is explained.

1 Introduction

Computer vision systems are made up of several stages in which they try to extract the information of the image. The stages and the purpose of its algorithms, depends on the application of the artificial vision system. Usually the systems are split in three main stages. The first stage is devoted to noise filtering, smoothing or contrast enhancement (it's also known as preprocessing or low level vision), the second stage is devoted to image segmentation, to split objects or regions according to their characteristics (intermediate level vision). Third stage involves the understanding of the scene (high level vision).

Uncertainty is present in every process of computer vision, therefore fuzzy techniques have been widely use in almost any of the processes. Extensions of fuzzy sets are not as specific as their counter-parts of fuzzy sets, but this lack of specificity makes them more realistic for some applications. Their advantage is that they allow us to express our uncertainty in identifying a particular membership function. This uncertainty is involved when extensions of fuzzy sets are processed, making results of the processing less specific but more reliable. Many authors based on this advantage proposed different image processing algorithms using extensions of fuzzy sets. This chapter presents a valuable review for the interested reader of the recent works using extensions of fuzzy sets in image processing. The chapter is divided as follows: first we recall the basics of the extensions of fuzzy sets, i.e. Type-2 fuzzy sets, Interval-valued fuzzy sets and Atanassov's Intuitionistic fuzzy sets. In sequent sections we review the methods

P. Melin et al. (Eds.): Bio-inspired Hybrid Intelligent Systems, SCI 256, pp. 3–32.
springerlink.com

proposed for noise removal (section 3), image enhancement (section 4), video deinterlacing (section 5), edge detection (section 6), segmentation (section 7), Stereo vision (section 8) and classification (section 9).

2 Extensions of Fuzzy Sets

From the beginning it was clear that fuzzy set theory [69] was an extraordinary tool for representing human knowledge. Nevertheless, Zadeh himself established (see [70]) that sometimes, in decision-making processes, knowledge is better represented by means of some generalizations of fuzzy sets. A key problem of representing the knowledge by means of Fuzzy sets is to choose the membership function which best represents such knowledge.

Sometimes, it is appropriate to represent the membership degree of each element to the fuzzy set by means of an interval. From these considerations arises the extension of fuzzy sets called *theory of interval-valued fuzzy sets*, that is, fuzzy sets such that the membership degree of each element of the fuzzy set is given by a closed subinterval of the interval $[0,1]$. Hence, not only vagueness (lack of sharp class boundaries), but also a feature of uncertainty (lack of information) can be addressed intuitively.

These sets were first introduced in the 1970s. In May 1975 Sambuc (see [54]) presented in his doctoral thesis the concept of an interval-valued fuzzy set named a Φ-fuzzy set. That same year, Zadeh [70] discussed the representation of type 2 fuzzy sets and its potential in approximate reasoning.

The concept of a *type 2 fuzzy set* was introduced by Zadeh [70] as a generalization of an ordinary fuzzy set. Type 2 fuzzy sets are characterized by a fuzzy membership function, that is, the membership value for each element of the set is itself a fuzzy set in $[0,1]$.

Formally, given the referential set U, a type 2 fuzzy set is defined as an object $\overline{\overline{A}}$ which has the following form:

$$\overline{\overline{A}} = \{(u,x,\mu_u(x))|u \in U, x \in [0,1]\},$$

where $x \in [0,1]$ is the primary membership degree of u and $\mu_u(x)$ is the secondary membership level, specific to a given pair (u,x).

One year later, Grattan-Guinness [34] established a definition of an interval-valued membership function. In that decade interval-valued fuzzy sets appeared in the literature in various guises and it was not until the 1980s, that the importance of these sets, as well as their name, was definitely established.

A particular case of a type 2 fuzzy set is an *interval type 2 fuzzy set* (see [41]–[42]). An interval type 2 fuzzy set $\overline{\overline{A}}$ in U is defined by

$$\overline{\overline{A}} = \{(u,A(u),\mu_u(x))|u \in U, A(u) \in L([0,1])\},$$

where $A(u)$ is a closed subinterval of $[0,1]$, and the function $\mu_u(x)$ represents the fuzzy set associated with the element $u \in U$ obtained when x covers the interval $[0,1]$; $\mu_u(x)$ is given in the following way:

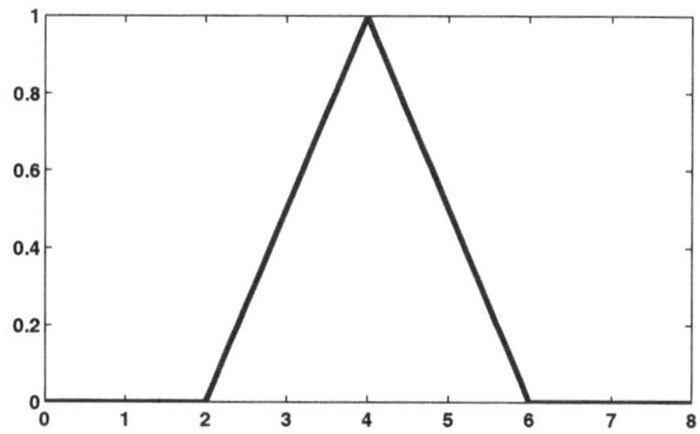

Fig. 1. Fuzzy membership function.

$$\mu_u(x) = \begin{cases} a \text{ if } \underline{A}(u) \leq x \leq \overline{A}(u) \\ 0 \text{ otherwise} \end{cases},$$

where $0 \leq a \leq 1$. It turns out that an interval type 2 fuzzy set is the same as an IVFS if we take $a = 1$.

Another important extension of fuzzy set theory is the theory of Atanassov's intuitionistic fuzzy sets ([1], [2]). Atanassov's intuitionistic fuzzy sets assign to each element of the universe not only a membership degree, but also a nonmembership degree, which is less than or equal to 1 minus the membership degree.

An Atanassov's intuitionistic fuzzy set (A-IFS) on U is a set

$$\hat{A} = \{(u, \mu_{\hat{A}}(u), \nu_{\hat{A}}(u)) | u \in U\},$$

where $\mu_{\hat{A}}(u) \in [0,1]$ denotes the membership degree and $\nu_{\hat{A}}(u) \in [0,1]$ the nonmembership degree of u in $\hat{A}$ and where, for all $u \in U$, $\mu_{\hat{A}}(u) + \nu_{\hat{A}}(u) \leq 1$.

In [1] Atanassov established that every Atanassov intuitionistic fuzzy set $\hat{A}$ on U can be represented by an interval-valued fuzzy set A given by

$$\begin{aligned} A : U &\to L([0,1]) \\ u &\to [\mu_{\hat{A}}(u), 1 - \nu_{\hat{A}}(u)], \qquad \text{for all } u \in U. \end{aligned}$$

Using this representation, Atanassov proposed in 1983 that Atanassov's intuitionistic fuzzy set theory was equivalent to the theory of interval-valued fuzzy sets. This equivalence was proven in 2003 by Deschrijver and Kerre [26]. Therefore, from a mathematical point of view, the results that we obtain for IVFSs are easily adaptable to A-IFSs and vice versa. Nevertheless, we need to point out that, conceptually, the two types of sets are totally different. This is made clear when applications of these sets are constructed (see [64]).

In 1993, Gau and Buehrer introduced the concept of *vague sets* [33]. Later, in 1996, it was proven that vague sets are in fact A-IFSs [6].

A compilation of the sets that are equivalent (from a mathematical point of view) to interval-valued fuzzy sets can be found in [27]. Two conclusions are drawn from this study:

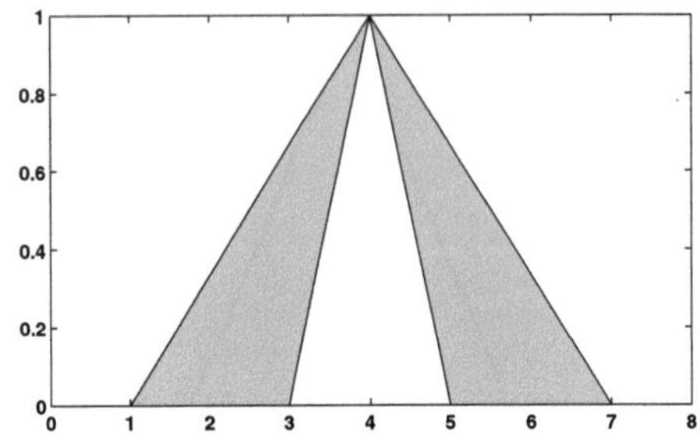

Fig. 2. Interval valued fuzzy membership function. Lower bound and upper bound

1.- Interval-valued fuzzy sets are equivalent to A-IFSs (and therefore vague sets), to grey sets (see [25]) and to L-fuzzy set in Goguen's sense with respect to a special lattice $L([0,1])$.
2.- IVFSs are a particular case of probabilistic sets (see [30]), of soft sets (see [3]), of Atanassov's interval-valued intuitionistic fuzzy sets and evidently of Type 2 fuzzy sets.

3 Noise Reduction

In this section we focus on gray scale images and the use of fuzzy logic theory extensions for noise removal.

Many applications of image processing perform image noise removal before any further processing such as segmentation, enhancement, edge detection or compression. Noise removal is a crucial task for most applications mainly for two reasons:

- Digital images are often corrupted by impulse noise during image acquisition, image transmission and image processing due to a number of imperfections present in these tasks environments.
- The noise can critically affect any image post-processing decisively compromising their performance.

For these reasons, noise removal is one of the most important steps for most image processing applications, regarding the overall performance of the applications and, is still a challenging problem in image processing.

When using fuzzy sets for noise removal, since their memberships are crisp values, it is not possible to know which one is the best membership function and, different membership functions will lead to different processing results. On the other hand, when using type-2 fuzzy sets, since their membership functions are also fuzzy, provides us with a more efficient way of dealing with uncertainty and, consequently, a more robust way of determining the membership functions.

In their work, Sun and Meng [57], make the assumption that, since impulse corruption is usually large compared with the strength of the signal, then the noise corrupted pixels have intensity values that are near the "saturated" values (lower and upper limits of the gray scale used). Based on this assumption, a inverted ladder

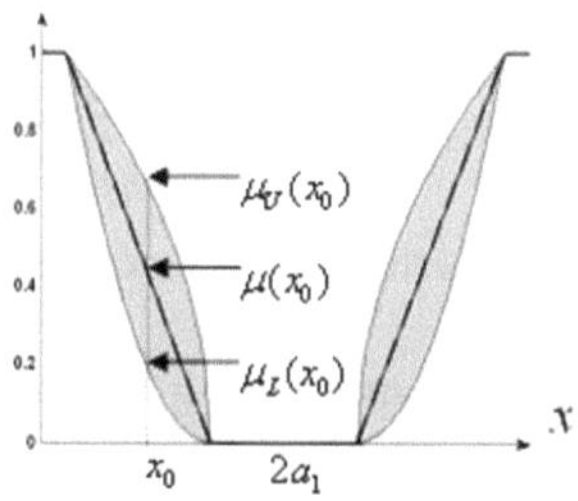

Fig. 3. Type-2 Fuzzy Set constructed from the initial ladder membership function

membership function is used as initial membership function from which the type-2 fuzzy set is constructed (Fig. 3).

After defining the inverted ladder membership function based on parameters a, a_1 and a_2, the lower and upper limits of the type-2 fuzzy set are obtained using the following equations:

$$\mu_U(x) = [\mu(x)]^{\frac{1}{\alpha}}$$
$$\mu_L(x) = [\mu(x)]^{\alpha}$$

where $\alpha \in (1,2]$.

After the construction of the type-2 fuzzy set, the probability of a pixel been corrupted by impulse noise (PPC) is represented by the centroid-type-reduction [67] of the set. This way, the larger PPC means the bigger probability of the pixel been corrupted and, the less PPC means the less probability. Finally, a threshold of corruption is established in such way that if the pixel's PPC is bigger than the threshold its restored value is the mean of the pixels intensities in the subimage ($n \times n$ filtering window centered at the pixel) and, if the pixel's PPC is less than the threshold it maintains its intensity.

Tulin, Basturk and Yuksel [62] proposed a type-2 fuzzy operator for detail preserving restoration of impulse noise corrupted images, that processes the pixels contained in a 3×3 filtering window and outputs the restored value of the window center pixel. The proposed operator is structure that combines four type-2 neuro-fuzzy filters, four defuzzifiers and a postprocessor. Each one of the type-2 NF filters are identical and accept the center pixel and two of its neighborhoods (processing the horizontal, vertical, diagonal and reverse diagonal pixel neighborhoods of the filtering window) and produces an output that represents the uncertainty interval (i.e., lower and upper bounds) for the center pixel restored value in the form of a type-1 interval. Each combination of the inputs of the filter and their associated type-2 interval Gaussian membership functions is represented by a rule in the form:

$$if(X_1 \in M_{i1})and(X_2 \in M_{i2})and(X_3 \in M_{i3}),then$$

$$R_i = k_{i1}X_1 + k_{i2}X_2 + k_{i3}X_3 + k_{i4}$$

There are N fuzzy rules in the rulebase where, R_i denotes the output of the ith rule (N rules, $i = 1, 2, \cdots N$) and M_{ij} denotes the ith Gaussian membership function of the jth input (3 inputs, $j = 1, 2, 3$) and is defined as follows:

$$M_{ij}(u) = \exp\left[-\frac{1}{2}(\frac{u-m_{ij}}{\sigma_{ij}})\right]$$

where m_{ij} and σ_{ij} are the mean and the standard deviation of the membership function, respectively.

Note that the membership functions M_{ij} (Fig. 4) are interval membership functions with their boundaries characterized by upper and lower Gaussian membership functions (see [62]).

The optimal values for the parameters of each one of the type-2 interval Gaussian membership functions are obtained by training using the least squares optimization algorithm. Finally, the output of each NF filter is the weighted average of the individual rules outputs.

The defuzzifier blocks convert the input fuzzy sets coming from the corresponding type-2 NF filters into a scalar value by performing centroid defuzzification (the centroid is the center of the type-1 interval fuzzy set).

The output of the defuzzifier blocks are four scalar values that are to be the candidates for the filtering window center pixel restored value. The postprocessor converts the four candidates into a single output value by discarding the lowest and the highest values and averaging the remaining two values.

In [62] Tulin, Basturk and Yuksel proposed an extension of this operator where the the original neighborhood topologies (horizontal, vertical, diagonal and reverse diagonal pixel neighborhoods) are extended to 28 possible neighborhood topologies corresponding to a filtering operator with 28 NF filters, 28 defuzzifiers and a postprocessor. However, authors emphasized that, for most filtering applications, one does not have to use all of these neighborhood topologies.

In this new operator, the NF filters and the defuzzifiers work in the same way as in the original operator. The way the new postprocessor produces the output value from the scalar values obtained at the outputs of the defuzzifiers is different from the original one. The postprocessor calculates the average of its inputs and truncates it to a 8-bit integer number, which is the output of the operator.

Wang, Chung, Hu and Wu [66] proposed a interval type-2 fuzzy filter for Gaussian noise suppression while maintaining the original structure of the image. Based on interval type-2 fuzzy sets is constructed a selective feedback fuzzy neural network (SFNN) suitable for image representation. In reality, this SFNN is a universal

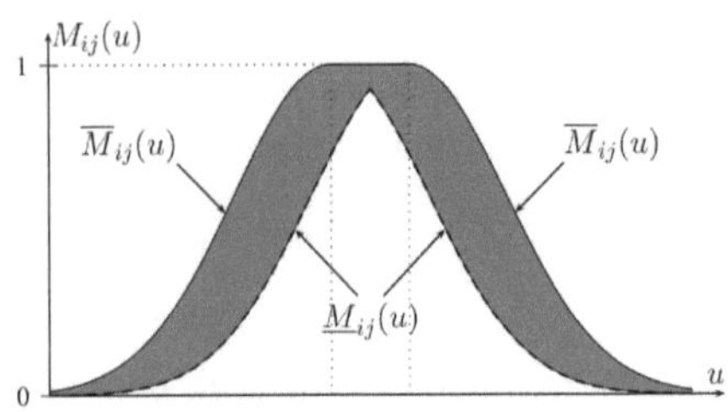

Fig. 4. Type-2 interval Gaussian membership function

approximator that works as a filter for Gaussian noise suppression that preserves the fine structure of the image from the theoretical viewpoint.

First, the image is fuzzified in using linguistic concepts in such way that the gray scale interval is equally divided in K_0 partitions. Each one of these partitions is described with a Gaussian shape and its membership function obtained by drawing all the Gaussians having mean and standard deviation.

The proposed SFNN has five components: two neurons in the input layer (since, the input is a 2-D digital image), three hidden layers and the output layer. Since, a digital image can be viewed as a continuous 2-D function and, according to the structure of the SFNN (see [66]) it can be used to approximate a continuous function, the SFNN is used to express a digital image and, a Gaussian noise filter is designed based on it. For each one of the K_0 partitions a optimal gray level is computed in a small operating window (mean of the pixels in the window) and, the window gray level which is nearest to the optimal gray level is selected as the optimal gray level in the operating window.

After the operating window slides over the whole image, the mean absolute error of the input image and the output image (filtered image) are calculated and if the difference between these two values is less than a small positive number the process stops; otherwise the gray levels of the input image are replaced by the gray levels of the output image and the process is restarted.

In [4], Bigand and Colot, use IVFS entropy to take into account the uncertainty present in the image noise removal process. Their motivation is to assess, and ultimately remove, the uncertainty of the membership values using the length of the interval in an IVFSs (the longer the interval, the more uncertainty).

For each pixel, the uncertainty of a precise FS is modeled by the closed intervals delimited by the upper ($\mu_U(x)$) and lower ($\mu_L(x)$) membership functions defined as follows:

$$\mu_U(x) = [\mu(x;g,\sigma)]^{\frac{1}{2}}$$
$$\mu_L(x) = [\mu(x;g,\sigma)]^2$$

where $\mu(x;g,\sigma)$ is a Gaussian fuzzy number defined as:

$$\mu(x;g,\sigma) = \exp\left[-\frac{1}{2}(\frac{x-g}{\sigma})^2\right]$$

Then, in order to be used as a criterion to automatically find fuzzy region width and thresholds for segmentation of noisy images, the entropy (which they wrongly called index of ultrafuzziness) of each one of the constructed IVFSs is calculated as follows:

$$\Gamma(x) = \frac{1}{M.N}\sum_{g=0}^{G-1}[h(x).(\mu_U(x) - \mu_L(x))]$$

where G is the number of gray levels of a $M \times N$ image.

This entropy is used to obtain a image homogram (in the same way as proposed by Cheng [24] using FSs) since, according to the authors, Γ represents the homogeneity distribution across intensities of the considered image. This homogram is used to find all major homogeneous regions while filtering noise in such way that, if a pixel belongs to one of these regions then the pixel is noise-free else pixel is noisy.

Finally, using a 3×3 filtering window, the median filter is applied to all the pixels identified as noisy pixels.

Discussion

Efficient noise removal in corrupted images is still a challenging problem in image processing mostly due to the imperfection/uncertainty inevitably present in noisy environments. The main idea of these works is to take into account the total amount of the imprecision/uncertainty present in the process of image noise filtering by means of the use of fuzzy sets extensions namely, interval-valued fuzzy sets and type-2 fuzzy sets.

All the presented filters inherit the advantages of fuzzy sets extensions theory, where an extra degree of fuzziness provides a more efficient way of dealing with uncertainty than with ordinary fuzzy sets.

Hence, using fuzzy sets extensions in noise filtering seems to bee a very promising idea that can lead to the design of efficient filtering operators.

4 Enhancement

In image enhancement, the main goal is to produce a new image that endows more accurate information for analysis than the original one. In this context, fuzzy logic extensions are used to represent and manipulate the uncertainty involved in the image enhancement process. Both type-2 fuzzy sets and A-IFSs approaches presented in this section, are able to model and minimize the effect that the uncertainty has in the image enhancement problem.

Image Enhancement using Type-2 Fuzzy Sets

In [28], Ensafi and Tizhoosh proposed a type-2 fuzzy image enhancement method based on extension of the locally adaptive fuzzy histogram hyperbolization method [59], improving its performance by extending it from a type-1 to a type-2 method where the additional third dimension of the type-2 sets gives more degrees of freedom for better representation of the uncertainties associated with the image.

First, based on the value of homogeneity μ_{Homo}, the image is divided into several sub-images.

$$\mu_{Homo} = \left(1 - \frac{g_{maxLocal} - g_{minLocal}}{g_{maxGlobal} - g_{minGlobal}}\right)^2$$

Setting the minimum and maximum local window sizes to 10 and 20 respectively and, using a fuzzy if-then-else rule, the local window size surrounding each supporting point is calculated.

The type-2 fuzzy set is obtained by blurring the type-1 fuzzy set defined by the membership function $\mu(g_{mn})$ of each gray level.

$$\mu(g_{mn}) = \frac{g_{mn} - g_{min}}{g_{max} - g_{min}}$$

Where g_{mn} represents the pixel gray level and, g_{min} and g_{max} represent the image minimum and maximum gray levels respectively.

The type-2 fuzzy set is constructed defining the upper and lower membership values using interval-valued fuzzy sets in the following way:

$$\mu_{UPPER}(x) = (\mu(x))^{0.5}$$

$$\mu_{LOWER}(x) = (\mu(x))^{2}$$

And, the proposed type-2 membership function is defined by:

$$\mu_{T2}(g_{mn}) = (\mu_{LOWER} \times \alpha) + (\mu_{UPPER} \times (1 - \alpha))$$

with,

$$\alpha = \frac{g_{Mean}}{L}$$

where, L is the number of gray levels and g_{Mean} the mean gray value of each sub-image.

Finally, using the μ_{T2} values, the new gray levels of the enhanced image are calculated using the following expression with $\beta = 1.1$:

$$g'_{mn} = \left(\frac{L \times 1}{e'^{1} \times 1}\right) \times \left[e'^{\mu(g_{mn})^{\beta}} - 1\right]$$

Image Enhancement using A-IFSs Entropy

In [65], Vlachos and Sergiadis studied the role of entropy in A-IFSs for contrast enhancement. Different concepts of entropy on A-IFss and their behavior are analyzed in the context of image enhancement.

In this work, image enhancement is regarded as a entropy optimization problem within an A-IFS image processing framework where, in the first stage the image is transferred into the fuzzy domain and sequentially into the A-IFS domain, where the proposed processing is performed. Finally, the inverse process is carried out in order to obtain the processed image in the gray-level domain.

Therefore, an intuitionistic fuzzification scheme for constructing an A-IFS for representing the image, based on entropy optimization, was proposed. First, the image ($N \times M$ pixels having L gray-levels) is represented in the fuzzy domain by A.

$$A = \{\langle g_{ij}, \mu_A(g_{ij})\rangle | g_{ij} \in 0, \times, L-1\}$$

with, $i \in \{1, \times, M\}$ and $j \in \{1, \times, N\}$.

A optimal derivation of a combination of membership and non-membership functions that model the image gray-levels is achieved by maximizing the intuitionistic fuzzy entropy of the image (i.e., maximum intuitionistic fuzzy entropy principle).

First, the membership function $\mu_A(g)$ of the fuzzified image is calculated.

$$\mu_A(g) = \frac{g - g_{min}}{g_{max} - g_{min}}$$

The A-IFS for representing the image,

$$\hat{A} = \{\langle g, \mu_{\hat{A}}(g;\lambda), \nu_{\hat{A}}(g;\lambda)\rangle | g \in 0, \times, L-1\}$$

is then constructed, by means of $\mu_A(g)$, using the following expressions:

$$\mu_{\hat{A}}(g,\lambda) = 1 - (1 - \mu_A(g))^{\lambda}$$

and

$$\nu_{\hat{A}}(g,\lambda) = (1 - \mu_A(g))^{\lambda(\lambda+1)}$$

where λ is obtained using an optimization criterion that is formulated as follows:

$$\lambda = \arg \max_{\lambda \geq 1} \{E(\hat{A};\lambda)\}$$

Where E is an entropy measure. Entropies proposed by Burillo and Bustince [5] and by Szmidt and Kacprzyk [56] where used.

The defuzzification is made using the "*maximum index of fuzziness intuitionistic defuzzification*" [64] by selecting a parameter α in the following way:

$$\alpha = \begin{cases} 0, & \text{if } \alpha' < 0 \\ \alpha', & \text{if } 0 \leqslant \alpha' \leqslant 1 \\ 1, & \text{if } \alpha' > 1 \end{cases}$$

where

$$\alpha' = \frac{\sum_{g=0}^{L-1} h_A(g)\pi_{\hat{A}}(g;\lambda)(1 - 2\mu_{\hat{A}}(g;\lambda))}{2\sum_{g=0}^{L-1} h_A(g)\pi_{\hat{A}}^2(g;\lambda)}$$

with h_A being the histogram of the fuzzified image A.

Finally, the new intensity gray levels g' are obtained through the expression:

$$g' = (L-1)\mu_{D\alpha(\hat{A})}(g)$$

where

$$\mu_{D\alpha(\hat{A})}(g) = \alpha + (1-\alpha)\mu_{\hat{A}}(g;\lambda) - \alpha\nu_{\hat{A}}(g;\lambda)$$

Discussion

In their work, Ensafi and Tizhoosh [28] demonstrated that using type-2 fuzzy logic a better image contrast enhancement is achieved than with its type-1 counterpart.

In Vlachos and Sergiadis approach [65], A-IFSs are used as a mathematical framework to deal with the vagueness present in a digital image by means of the A-IFS extra degree of freedom that allows a flexible modeling of imprecise and/or imperfect information present in images, better than classical fuzzy sets. They concluded that the different notions of intuitionistic fuzzy entropy used [5, 56] treat images in different ways, making the selection of the appropriate entropy measure to be application-dependent.

5 Video De-interlacing

Deinterlacing has been widely used in television and electronic video cameras industry. Its great contribution is to double the refreshing rate without increasing the bandwidth. Deinterlacing converts each field into a frame, so the number of pictures per second remains constant while the number of lines per picture is doubled [37]. Deinterlacing has some problems as flickering because of the intrinsic nature of the interlaced scanning process. Deinterlacing methods can be divided in three categories.

- Spatial domain methods: pixels are interpolated only from the pixels available on that field.
- Temporal domain methods: pixels are calculated based on the pixels received before.
- Spatio-temporal domain methods: work with pixels received in the same field and the ones received before (usually with motion compensation).

In [38] Jeon et. al propose a type-2 fuzzy deinterlacing system (T2DF). This system has three sections: a Fuzzifier, a Fuzzy Inference Engine and a Defuzzifier. The Fuzzifier inputs the neighbour pixels of a missing one and outputs eighteen values representing the luminance difference. These values are combined in spatial domain and temporal domain, the displacement in horizontal direction and the edge direction (45°, 90° or 135°).

In the Fuzzy Inference Engine the input corresponds to those obtained values. Through eighteen fuzzy filters a type-2 interval fuzzy set is obtained representing the uncertainty interval for the restored value of the deinterlaced pixel. This fuzzy set is combined in six weighting factors which output six scalar values.

Those last values, considered candidate deinterlaced pixel values, are the input for the Defuzzifier. By a weighted average processor, the input is transformed into a single scalar value. The output obtained by the Defuzzifier, which is the global T2DF output, represents the restored value for the center pixel of the filtering window.

In [39] Jeon et al. propose a two-stage algorithm. The first stage is a spatio-temporal fuzzy approach to video deinterlacing while the second one is a rough sets-assisted optimization. The fuzzy approach's main goal is to create a fuzzy if-then rules set. For this purpose the edge directions are calculated in space and time. The difference values are labelled by linguistic descriptions. From this, they create four if-then rules to represent if the pixel is in an edge (two rules), in a monotonic slope or in a peak. Once the pixel environment is identified, it can be well interpolated. The optimization algorithm combines five deinterlacing methods to evaluate the missing pixels. This methods are Bob, Weave, ELA, STELA and FD (see [29]). It is organized as a decision table, where the minimal gradient values are condition attributes (input), and the output is a decision attribute which represents the method that is selected. This decision table is reduced maintaining the results to a less time-consuming execution.

The main advantage of the use of fuzzy extensions in deinterlacing methods is to handle uncertain information effectively. On the other hand, this improvement has a high computational cost. Besides, the parameters for the fuzzy membership functions must be determined experimentally, which is also a limitation.

6 Edge Detection

In the ideal case, the result of applying an edge detector to an image may lead to a set of connected curves that indicate the boundaries of objects, the boundaries of surface markings as well curves that correspond to discontinuities in surface orientation. Thus, applying an edge detector to an image may significantly reduce the amount of data to be processed and may therefore filter out information that may be regarded as less relevant, while preserving the important structural properties of an image.

One of the seminal works of edge detection was Canny's [21]. Its aim was to discover the optimal edge detection algorithm. In this situation, an optimal edge detector means:

- Good detection - the algorithm should mark as many real edges in the image as possible.
- Good localization - edges marked should be as close as possible to the edge in the real image.
- Minimal response - a given edge in the image should only be marked once, and where possible, image noise should not create false edges.

Some fuzzy approaches to edge detection already exist and perform quite well (see [35]). Why then work with an extension of fuzzy sets? A classical definition of what an edge within an image should be is: *a significantly change in the intensity of adjacent neighboring pixels*. To state a specific threshold on how large the intensity change between two neighboring pixels must be, is not a simple task and obviously depends on the scene, illumination etc. So it's clear that extensions of fuzzy sets can be used to deal with this uncertain concept.

In the literature of extensions of fuzzy sets applied to edge detection there exist three different approaches. In the first one the main idea is to assign each pixel of the image with an interval, and then measuring its entropy decide if it's edge or not.

Fig. 5. Example of edge detection

In the second approach a interval type fuzzy system is used to classify pixels and, in the third approach AIFSs are used to deal with the uncertainty of edge pattern matching.

First approach [18]: Consider the fact that edge detection techniques attempt to find pixels whose intensity (gray level) is very different from those of its neighbors. An element of f (image) belongs to an edge if there is a big enough difference between its intensity and its neighbors intensities. (Notice that this definition is intentionally fuzzy in its own right). The method begins assigning an IVFS to each matrix f and therefore each element has associated an interval as membership degree. The lower and upper bounds of this interval are determined by the concepts of tn-processing and sn-processing.

Definition 1. ([16]) Consider a matrix $f \in \mathscr{M}$, any two t-norms T_1 and T_2 in $[0,1]$, and a positive integer n less than or equal to $\frac{N-1}{2}$ and $\frac{M-1}{2}$. We define the *tn-processing* of f as follows:

$$g^n_{T_1,T_2} : \mathscr{M} \to \mathscr{M}_n \text{ given by}$$
$$g^n_{T_1,T_2}(f(x,y)) = \underset{\substack{i=-n\\ j=-n}}{\overset{n}{T_1}} \left(T_2(f(x-i,y-j),f(x,y))\right)$$
$$\text{with } n \leq x \leq N-(n+1), n \leq y \leq M-(n+1)$$

In this case it is said that we use a submatrix of order $(2n+1)\times(2n+1)$.

Definition 2. ([16]) Let n be a whole number greater than zero. We define the *IVn matrix* associated with $f \in \mathscr{M}$ as the interval-valued fuzzy set G^n given by

$$G^n = \left\{\left((x,y), G^n(x,y) = \left[g^n_{T_1,T_2}(f(x,y)), g^n_{S_1,S_2}(f(x,y))\right] \in L([0,1])\right) | x \in X, y \in Y\right\}$$

being $g^n_{T_1,T_2}$ and $g^n_{S_1,S_2}$ the tn-processing and the sn-processing given by Definition 1.

Obviously, we can also associate with each matrix f the following interval-valued fuzzy set $\mathbf{f}$: $\mathbf{f} = \{((x,y), \mathbf{f}(x,y) = [f(x,y), f(x,y)] \in L([0,1])) | x \in X, y \in Y\}$.

The following definition associates with the IVFS G^n a fuzzy set whose membership function is the length of the intervals in G^n.

Definition 3. ([16]) Given a matrix $f \in \mathscr{M}$ and its corresponding G^n. We call W-matrix of f, a new matrix obtained by assigning to each of its elements the corresponding interval length of G^n. It is denoted as $W(G^n)$. Therefore
$W(G^n) = \left\{\left((x,y), W(G^n)(x,y) = g^n_{S_1,S_2}(f(x,y)) - g^n_{T_1,T_2}(f(x,y))\right) | (x,y) \in X \times Y\right\} \in FSs(X \times Y)$.

Every matrix f is thus associated with an interval-valued fuzzy set G^n and a fuzzy set $W(G^n)$.

The normalized entropy of G^n is given by the following expression:

$$\mathscr{E}_{\mathscr{N}}(G^n) = \frac{\sum\limits_{\substack{n \leq x \leq N-(n+1)\\ n \leq y \leq M-(n+1)}} g^n_{S_1,S_2}(f(x,y)) - g^n_{T_1,T_2}(f(x,y))}{(N-2\cdot n)\times(M-2\cdot n)} \tag{1}$$

It is logical to relate those elements of G^n whose membership degree intervals are large to the location of the edge. This fact leads us to establish the following:

The normalized entropy of G^n establishes the average length of the intervals that represent the membership function of the elements. A large entropy implies that the intervals are large, meaning that the difference between $g^n_{T_1,T_2}$ and $g^n_{S_1,S_2}$ tends to be large for each element. In such a matrix there are many changes in intensity, and a higher proportion of elements belong to an edge. Conversely, a small entropy implies that few elements of the matrix belong to an edge. The normalized entropy of G^n therefore indicates the *proportion of elements* that are part of an edge.

Definition 4. We say that an element (x,y) belongs to the bright zone edge if it satisfies the following two items:

(i) $f(x,y) \geq \frac{g^n_{\vee,\vee}(f(x,y))+g^n_{\wedge,\wedge}(f(x,y))}{2} = K_{0.5}(G^n(x,y))$, and
(ii) its interval has sufficient length.

To identify elements of G^n with a long enough interval to belong to an edge, we start by considering two intensity values p and q whose values are yet to be determined. For the moment, we simply require $p \leq q$ and $p,q \in [0,1]$ (in a binary image, we would have $p = 0$ and $q = 1$). From these two values we base on the following rule to obtain the edges:

(a) *If $W(G^n(x,y)) \geq q$, then the element belongs to an edge.*

(b) *If $p \leq W(G^n(x,y)) < q$, then we need a way of distinguishing elements that belong to the edge from those that do not belong.*

(c) *If $W(G^n(x,y)) < p$, then the element (x,y) generally does not belong to the edge.* Such elements can be considered, however, if very few elements satisfy conditions (a) and (b).

From previous rule is proposed the following algorithm, in which relevant pixels are added to the edge by means of different actions (please see [18]).

(BZ1) *Calculate the IVn matrix G^n and its associated W-matrix (by means of $g^n_{\wedge,\wedge}$ and $g^n_{\vee,\vee}$).*
(BZ2) *Calculate p and q, then construct the sets G^n_p, $G^n_{p,q}$ and G^n_q.*
(BZ3) *Calculate the entropies $\mathcal{E}_{\mathcal{N}}(G^n_p)$, $\mathcal{E}_{\mathcal{N}}(G^n_{p,q})$ and $\mathcal{E}_{\mathcal{N}}(G^n_q)$.*
(BZ4) *Calculate $\mathbf{T}_{\wedge,\wedge}(\mathbf{f}, G^n)$.*
(BZ5) *Calculate $\mathcal{E}_{\mathcal{N}}(\mathbf{T}_{\wedge,\wedge}(\mathbf{f}, G^n)^n_p)$, $\mathcal{E}_{\mathcal{N}}(\mathbf{T}_{\wedge,\wedge}(\mathbf{f}, G^n)^n_{p,q})$ and $\mathcal{E}_{\mathcal{N}}(\mathbf{T}_{\wedge,\wedge}(\mathbf{f}, G^n)^n_q)$.*
(BZ6) *Execute **(Action1)**, **(Action2)** or **(Action3)**.*
(BZ7) *Sum the binary images obtained in (BZ6).*
(BZ8) *Clean and thin the lines.*

Second approach [44, 45]: The goal is to design a system which makes it easier to include edges in low contrast regions, but which does not favor false edges by effect of noise. Because of this specifications the authors design an Interval type 2 fuzzy system.

The system has 4 inputs and one output that is the degree of edginess of each pixel.

The input variables are the gradients with respect to x-axis and y-axis, to which they call DH and DV respectively. The other two inputs are the pixels filtered when convolute two masks to the original image. One is a high-pass filter and the other a low-pass filter. The high-pass filter HP detects the contrast of the image to guarantee the border detection in relative low contrast regions. The low-pass filter M allow to detect image pixels belonging to regions of the input were the mean gray level is lower. These regions are proportionally more affected by noise, supposed uniformly distributed over the whole image.

Seven interval valued fuzzy rules allow to evaluate the input variables, so that the obtained image displays the edges of the image in color near white (HIGH tone), whereas the background was in tones near black (tone LOW).

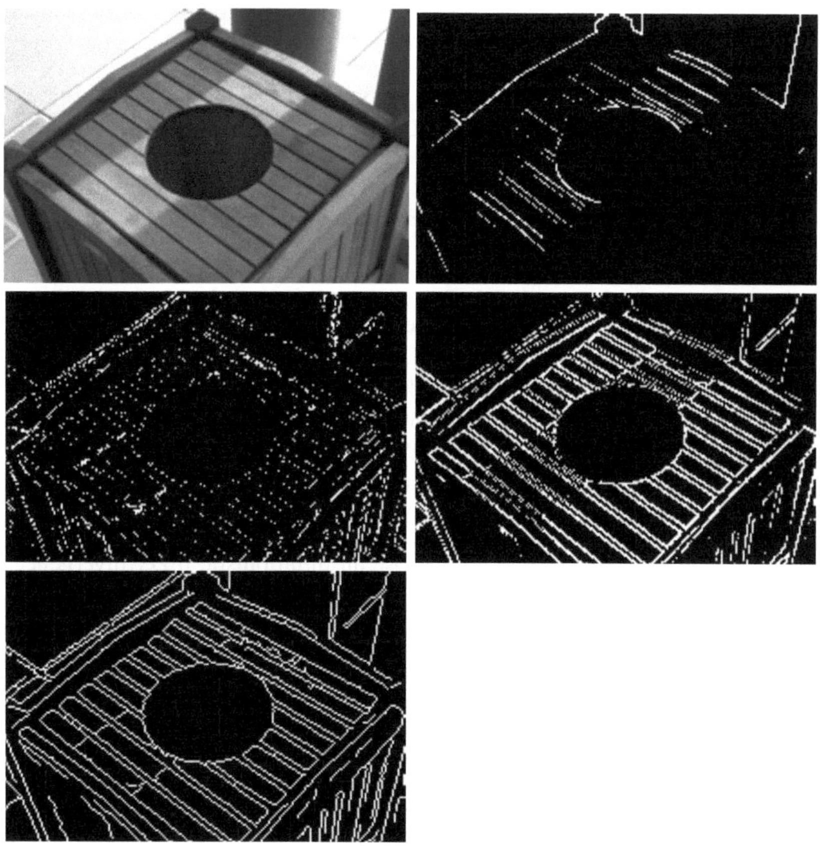

Fig. 6. Different edges that are obtained in the process

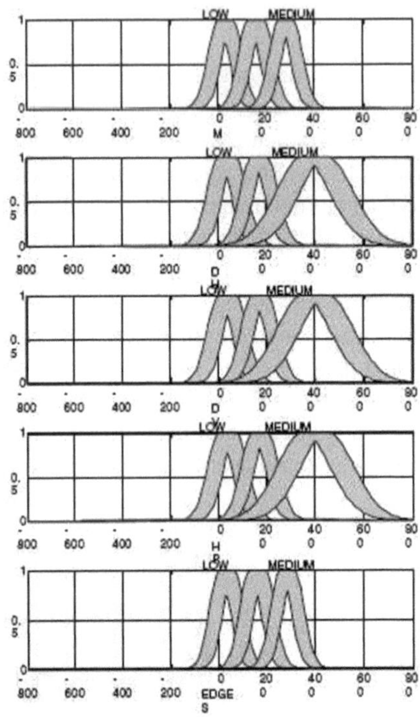

Fig. 7. Interval valued membership functions designed for the input variables and the output variable: Degree of "edginess".

1. If (DH is LOW) and (DV is LOW) then (EDGES is LOW)
2. If (DH is MEDIUM) and (DV is MEDIUM) then (EDGES is HIGH)
3. If (DH is HIGH) and (DV is HIGH) then (EDGES is HIGH)
4. If (DH is MEDIUM) and (HP is LOW) then (EDGES is HIGH)
5. If (DV is MEDIUM) and (HP is LOW) then (EDGES is HIGH)
6. If (M is LOW) and (DV is MEDIUM) then (EDGES is LOW)
7. If (M is LOW) and (DH is MEDIUM) then (EDGES is LOW)

Third Approach [22]: The authors propose an intuitionistic fuzzy divergence to deal with the uncertainty in an edge pattern matching scheme. The intuitionistic fuzzy set takes into account the uncertainty in assignment of membership degree known as hesitation degree.

The main idea behind template edge matching is to detect tipical intensity distributions that are usually in edges. The authors propose the following algorithm to deal with uncertainty present in the matching process by means of an intuitionistic divergence.

- Step 1. Form 16 edge-detected templates.
- Step 2. Apply the edge templates over the image by placing the center of each template at each point (i,j) over the normalized image.

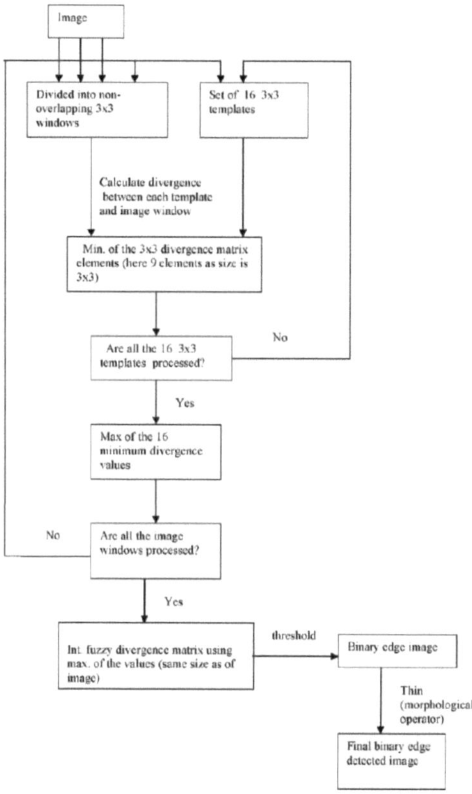

Fig. 8. Chaira's algorithm

- Step 3. Calculate the intuitionistic fuzzy divergence (IFD) between each elements of each template and the image window and choose the minimum IFD value.
- Step 4. Choose the maximum of all the 16 minimum intuitionistic fuzzy divergence values.
- Step 5. Position the maximum value at the point where the template was centered over the image.
- Step 6. For all the pixel positions, the maxmin value has been selected and positioned.
- Step 7. A new intuitionistic divergence matrix has been formed.
- Step 8. Threshold the intuitionistic divergence matrix and thin.
- Step 9. An edge-detected image is obtained.

Experimental studies reveal that, for edge detection the result is completely dependent on the selection of hesitation constant and thereby by the hesitation degree (also called the intuitionistic fuzzy index). The intuitionistic method detects the dominant edges clearly, while removing the unwanted edges.

7 Segmentation

Image segmentation is the first step in image analysis and pattern recognition. It is a critical and essential component of image analysis and/or pattern recognition system and is one of the most difficult tasks in image processing, that can determine the quality of the final result of the system.

The goal of image segmentation is the partition of an image in different areas or regions.

Definition 5. Segmentation is grouping pixels into regions such that

1. $\cup_{i=1}^{k} P_i =$ Entire image ($\{P_i\}$ is an exhaustive partitioning).
2. $P_i \cap P_j = 0, i \neq j$ ($\{P_i\}$ is an exclusive partitioning).
3. Each region P_i satisfies a predicate; that is, all points of the partition have some common property.
4. Pixels belonging to adjacent regions, when taken jointly, do not satisfy the predicate.

There exist three different approaches using fuzzy methods:

- Histogram thresholding.
- Feature space clustering.
- Rule based systems.

There exist different works that use extensions of fuzzy sets within the three approaches. The most commonly studied method is thresholding with extensions of fuzzy sets. The first work on this topic was made by Tizhoosh [58].

One of the earlier papers of fuzzy thresholding is [52] in 1983. The idea behind the fuzzy thresholding is to first transfer the selected image feature into a fuzzy subset by means of a proper membership function and then select and optimize a global or local fuzzy measure to attain the goal of image segmentation.

In [52] the authors used the S-function to fuzzify the image. They minimize the entropy in such a way that the final segmented image is the one which has less doubtful pixels.

Said membership functions represent the brightness set within the image. The basic idea of using this membership function is that, if we take the value of a parameter as the threshold value, the dark pixels should have low membership degrees, and on the contrary, brighter pixels should have high membership degrees. The pixels with membership function near 0.5 should be the ones that are not clearly classified. Therefore the set with less entropy is the set with less amount of pixels with uncertain membership (around 0.5).

In 2005 Tizhoosh [58] presented a paper that uses Interval type 2 fuzzy sets in image thresholding (we must point out that he tries to use type 2 fuzzy sets, however in the paper he only uses Interval Type 2 fuzzy sets). His study is based on the modification of the classical fuzzy algorithm of Huang and Wang [36], so that he applies an α factor as interval generator to the membership function. Starting from a membership function, Tizhoosh obtains an interval valued fuzzy set that "contains" different membership functions and is useful for finding the threshold of an image.

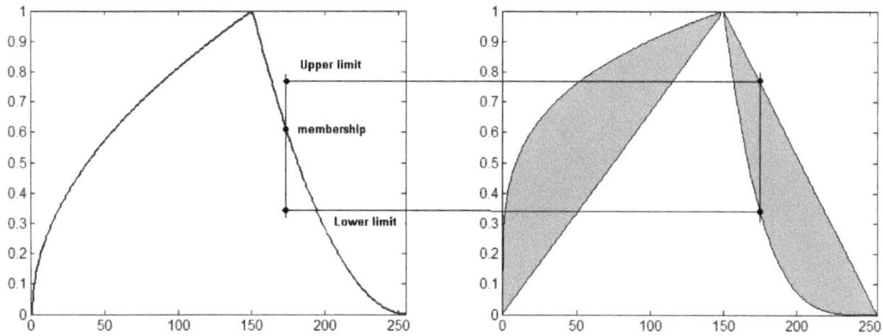

Fig. 9. Threshold as a fuzzy number used by Tizhoosh. Transformation of a FS into a IVFS.

He assigns each intensity with the following interval of membership, μ_U and μ_L being the upper and lower membership degrees:

$$\mu_L(g) = \mu(g)^\alpha \qquad (2)$$
$$\mu_U(g) = \mu(g)^{1/\alpha}$$

With $\alpha \in (1,\infty)$, and therefore $0 \leq \mu_L(q) \leq \mu_U(q) \leq 1$. Sometimes parameter α can be interpreted as linguistic edges.

Tizhoosh's idea for proposing his algorithm is to *"remove the uncertainty of membership values by using type II fuzzy sets"*.

Vlachos and Sergiaidis [64] also propose a modification of Tizhoosh's algorithm, using Atanassov's intuitionistic fuzzy sets (see [1]). Their basis are membership functions similar to Huang's but, instead of minimizing the entropy, the algorithm minimizes the divergence with set $\hat{\mathbf{1}}$ (see [23]). The structure of the intuitionistic algorithm is the same as Tizhoosh's. The construction of the intuitionistic fuzzy sets is done in the following way:

$$\mu_{\hat{A}}(g,t) = \lambda \mu_A(g,t) \qquad (3)$$
$$\nu_{\hat{A}}(g,t) = (\hat{1} - \lambda \mu_A(g,t))^\lambda$$

With $\lambda \in [0,1]$, being $\hat{A}$ an intuitionistic fuzzy set, and the divergence:

$$D_{IFS}(\hat{A},\hat{1},t) = \sum_{g=0}^{L-1} h_A(g) \left(\mu_{\hat{A}(g,t)} ln \frac{2\mu_{\hat{A}}(g,t)}{1+\mu_{\hat{A}}(g,t)} + \nu_{\hat{A}}(g,t) ln2 + ln \frac{2}{1+\mu_{\hat{A}}(g,t)} \right) \qquad (4)$$

Tizhooh's algorithm is applied directly to color segmentation using RGB in [48]. Moreover in [68] it's used to segment color image skin lesions.

But there exist an improvement of Tizhoosh algorithm, that arises from the selection of the membership functions. It was proved, that the membership functions that best represent the image are the ones used in [32, 36], that represent how similar the intensity of each pixel is to the mean of the intensities of the object or to the mean of the intensities of the background. By defining the functions in this way, the set with

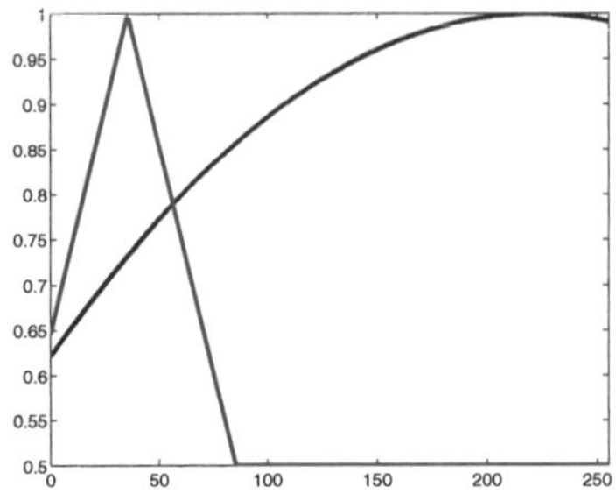

Fig. 10. Two different membership functions to represent the background and the object with t=150

lowest entropy is the set that contains the greatest number of pixels around the mean of the intensities of the background and the mean of the intensities of the object.

Starting form the idea of obtaining the uncertainty from the information given by the user, we have proposed several approximations using A-IFS [13, 14] (where the key point is to calculate the intuitionistic index) and also using interval valued fuzzy sets [11] (where the key point is to calculate the lengths of the intervals). These works lead us to introduce the concept of an ignorance function to try to model the lack of knowledge from which experts may suffer when determining the membership degrees of some pixels of an image Q to the fuzzy set representing the background of the image, Q_B, and to the fuzzy set representing the object in the image, Q_O.

The classical fuzzy thresholding algorithm is modified due to the user should pick two functions, one to represent the background and another one to represent the object. We have chosen this representation since, in this way, the expert is able to get a better representation of the pixels for which he is not sure of their membership to the object or the background. In figure 10 we show two membership functions, one to represent the background and the other to represent the object.

As we have already said in the previous paragraph, we are going to represent the images by means of two different fuzzy sets. For this reason, in our proposed algorithm we introduce the concept of *ignorance function* G_u. Such functions are a way to represent the user's ignorance for choosing the two membership functions used to represent the image (object and background). Therefore, in our algorithm we will associate to each pixel three numerical values:

- A value for representing its membership to the background, which we will interpret as the expert's knowledge of the membership of the pixel to the background.
- A value for representing its belongingness to the object, which we will interpret as the expert's knowledge of the membership of the pixel to the object.
- A value for representing the expert's ignorance of the membership of the pixels to the background or to the object. This ignorance hinders the expert from making an exact construction of the membership functions described in the first two items and therefore it also hinders the proper construction of step (a) of the fuzzy algorithm. The lower the value of ignorance is, the better the membership function chosen to represent the membership of that pixel to the background and the one

chosen to represent the membership to the object will be. Evidently, there will be pixels of the image for which the expert will know exactly their membership to the background or to the object but there will also be pixels for which the expert is not able to determine if they belong to the background or to the object.

Under these conditions, if the value of the function of ignorance (G_u) for a certain pixel is zero, it means that the expert is positively sure about the belongingness of the pixel to the background or to the object. However, if the expert does not know at all whether the pixel belongs to the background or to the object he must represent its membership to both with the value 0.5, and under these conditions we can say that the expert has *total* ignorance regarding the membership of the pixel to the background and the membership of the same pixel to the object.

In [19] a methodology is proposed to construct Ignorance functions. In the following example we show an ignorance function that can be constructed by means of the t-norm minimum.

Example 1. Using the t-norm $T = T_M$

$$G_u(x,y) = \begin{cases} 2 \cdot min(1-x, 1-y) \text{ if } min(1-x,1-y) \leq 0.5 \\ \frac{1}{2 \cdot min(1-x,1-y)} \text{ otherwise} \end{cases}$$

is a continuous ignorance function.

Also Ignorance functions can be constructed from other t-norms as the product or other functions like the geometric mean that are not t-norms (please see [19]).

In the following proposition we can see how we can construct IVFS by means of the ignorance function.

Proposition 1. *Let $\tilde{Q}_B$ and $\tilde{Q}_O$ be the fuzzy sets associated to the background and the object of the image built by an expert. Let G_u be the ignorance function associated to the construction of the said fuzzy sets. Under these conditions, if we define*

$$\begin{aligned} &\Phi : \mathscr{F}\mathscr{S}s(U) \times \mathscr{F}\mathscr{S}s(U) \longrightarrow \mathscr{I}\mathscr{V}\mathscr{F}\mathscr{S}s(U) \text{ given by} \\ &\Phi(\tilde{Q}_B, \tilde{Q}_O) = \{(u, [M_L(u), M_U(u)]) | u \in U\} \text{ such that} \\ &[M_L(u), M_U(u)] = [G_u(0.5, 0.5) - G_u(\mu_{\tilde{Q}_B}(u), \mu_{\tilde{Q}_O}(u)), G_u(0.5, 0.5)], \end{aligned}$$

then for each membership $[M_L(u), M_U(u)]$ defined by Φ the following equation holds:

$$W([M_L(u), M_U(u)]) = G_u(\mu_{\tilde{Q}_B}(u), \mu_{\tilde{Q}_O}(u))$$

To sum up, the goal of this algorithms is: *To prove that the result of the algorithm we propose, which uses interval valued fuzzy sets built from two membership functions, improves the result provided by the fuzzy algorithm.*

In the experiments we intend to check the following two items to prove that the result of the algorithm that we propose, which uses interval valued fuzzy sets constructed from two membership functions and an ignorance function, improves the result provided by the fuzzy algorithm:

1. The solution provided by the IVFS algorithm should always be better than the solution provided by the generalized fuzzy algorithm when the membership functions have been chosen in a bad way.

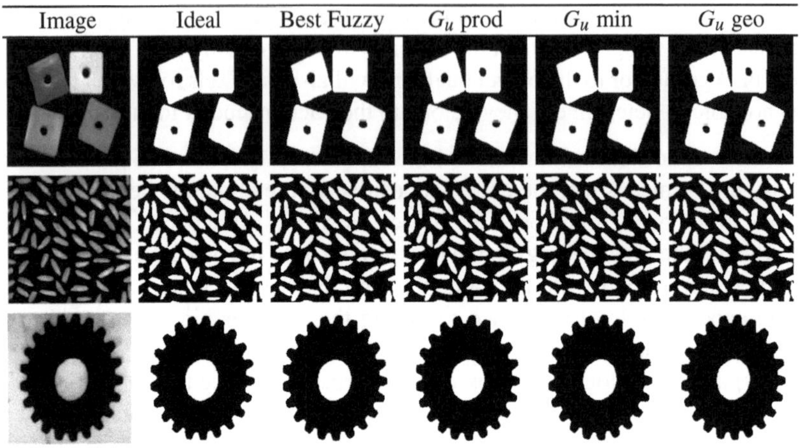

Fig. 11. Best binary images obtained with different methods.

2. The solution provided by the IVFS algorithm should should be equal to or better than the solution provided by the generalized fuzzy algorithm if the chosen membership functions represent correctly the background and the object.

In [19] it's proved that for certain images images both items are satisfied, and for special type of images (ultrasound images) only the first item is satisfied.

In figure 11 we show a table with the best segmentations obtained with the generalized fuzzy algorithm (using two different membership functions), and the IVFS algorithm when using the product, the minimum and the geometric mean as ignorance G_u functions.

Clustering: A very common method for segmenting images is clustering. The most studied algorithm for this purpose is the Fuzzy Cluster Means (FCM), which aims to find the most characteristic point of each cluster, considered its centroid, and the membership degree of every object to each cluster. Some authors have adapted this algorithm to type-2 fuzzy sets. In [40] Hwang et al. try to define and manage the uncertainty of fuzzifier m in FCM. They define the lower and upper interval memberships using two different values of m. To manage appropriately the uncertainty defined in an interval type-2 fuzzy set through all steps of FCM, they update cluster centers employing type reduction and defuzzification methods using type-2 fuzzy operations.

Fuzzy rule based systems: Data driven methods have also used in image segmentation. Starting from man made segmentations as ground truth data, machine learning algorithms have been used to create intellingent systems devoted to segment similar images as the trainig ones. Neural networks are the most common, and also fuzzy rule based systems have provided good results.

In [20] we introduce an application of interval-valued systems to the segmentation of prostate ultrasound images. The system classifies each pixel as prostate or background. The input variables are the values of each pixel in different processed images as proximity, edginess and enhanced image. The system has 20 rules and is trained with ideal images segmented by an expert.

8 Stereo Vision

Since images are 2D projections of 3D scenes, depth recovery is of paramount importance in image understanding in terms of its 3D scene structure, with many applications in passive navigation, cartography, surveillance systems, industrial robotics, etc.

One of the techniques used to get depth perception is through the use of stereo analysis. Stereo analysis provides a more direct quantitative depth evaluation than techniques such as shape from shading, and its being passive makes it more applicable than active range finding imagery by laser or radar.

In this case, depth can be evaluated from the difference in the left and right images location a 3D scene point location takes. The stereo matching problem consists in the determination of this difference, called disparity, of a certain scene feature.

We can see in Fig. 12 that it involves the identification of the same physical points in both images. Assuming that the epipolar constraint stands, the same physical points (x_1, x_2, x_3) in the left and right images have different abscissa in the image coordinate system. The determination of these corresponding points can be difficult because it may be unclear which point in the left image corresponds to which point in the right image, due to occlusion, perspective distortion, different lighting intensity, reflections, shadows, repetitive pattern, sensory noise, etc.

There are many algorithms for solving the general stereo matching problem [55], based on disparate approaches including fuzzy logic [60]. The use of techniques using fuzzy sets extensions is important in establishing robust approaches to the correspondence problem, allowing for the incorporation of the associated uncertainty.

There are two similar approaches to the stereo matching problem (or to the underlying correspondence problem) based on extensions of Fuzzy Sets: one uses

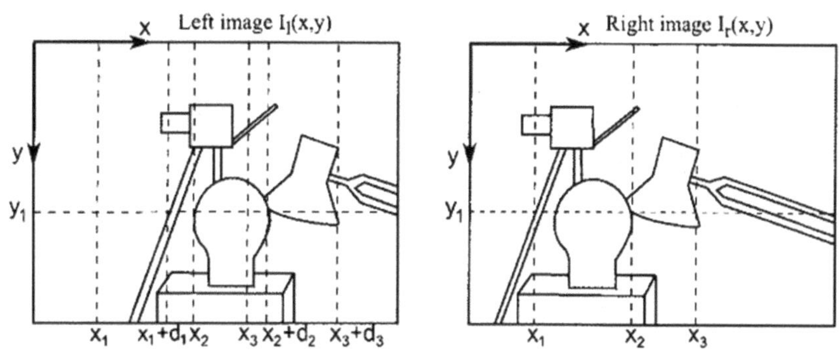

Fig. 12. The stereo matching problem

Atanassov's Intuitionistic Fuzzy Sets (A-IFSs) [47], while the other uses Interval-Valued Fuzzy Sets (IVFSs) [17].

Both algorithms have essentially the following steps:

A Fuzzification of the right and left images.
B Construction of the extended FSs associated with the left and the right images.
C Construction of the disparity matrix from the two extended FSs

Stereo Matching using A-IFSs

For the fuzzification of the images Nieradka [47] uses information and fuzziness to determine the membership S-function. The shape of the S-function should contain enough information about the fuzzy set. This is done defining an index of fuzziness over the image histogram:

$$\gamma_{linear}(I) = \frac{2}{MN} \sum_{g=0}^{L-1} h^c(g).\min[\mu_I(g), \overline{\mu}_{\overline{I}}(g)]$$

where I is an image of size $M \times N$ with L gray levels, $h^c(g)$ is the image histogram and $\mu_{\overline{I}} = 1 - \mu_O(g)$. Also it should be the most informative as possible, using entropy as a measure of information:

$$H = -\sum_{i=1}^{n} p_i \log(p_i) + \sum_{i=1}^{n} p_i E(\mu_i)$$

$$E(\mu_i) = -\mu_i \log(\mu_i) - (1-\mu_i)\log(1-\mu_i)$$

where $p_1, p_2, \cdots, p_n$ are the probabilities of n randomly occurring events $x_1, x_2, , x_n$. The average value for parameters a, b and c is then chosen (whereas maximizing a, b and c for entropy and for fuzziness index is considered), as the best choice for a balance between information and fuzziness.

Construction of the A-IFSs associated with the left and the right images

A hesitancy component is determined (in order to take into account quantization noise that can somehow affect the previous membership function evaluation), with the use of the image fuzzy histogram $h^f(g)$:

$$h^f(g) = \|(i,j)|\mu_g(g_{ij}); i = 1, \cdots, M, j = 1, \cdots, N\|$$

The fuzzy number $\mu_g(x)$ is defined as follows:

$$\mu_g(x) = \max\left(0, 1 - \frac{|x-g|}{p}\right)$$

where p controls the fuzzy number spread. The hesitancy is then defined as proportional to the normalized absolute difference between the crisp and the fuzzy histograms:

$$\pi_{\hat{A}}(g) \propto \frac{\widetilde{h}^c(g) - \widetilde{h}^f(g)}{\max_g ||\widetilde{h}^c(g) - \widetilde{h}^f(g)||}$$

$$\pi_{\hat{A}}(g) = (1 - \mu_A(g)) \frac{\widetilde{h}^c(g) - \widetilde{h}^f(g)}{\max |\widetilde{h}^c(g) - \widetilde{h}^f(g)|}(1 - k\Delta r)$$

where $\widetilde{h}^c(g)$ and $\widetilde{h}^f(g)$ are the normalized crisp and fuzzy histograms respectively, , and $k \in (0,1)$ controls the influence of the dynamic range of the image.

The matching score is calculated, using the following similarity measure between A-IFSs A and B:

$$SM^p_{\hat{A}\hat{B}} = 1 - \frac{1}{\sqrt[p]{n}} \sqrt[p]{\sum_{i=1}^{n} (\varphi_{\mu\hat{A}\hat{B}}(x) - \varphi_{\nu\hat{A}\hat{B}}(x))^p}$$

Stereo Matching using IVFSs

The methodology proposed by Bustince [17] does the fuzzification of the images using the membership function associated with the best image segmentation.

$$\mu_{I_t}(q) = \begin{cases} \varphi(REF(q, m_b(t))) & \text{if} \quad q \leq t \\ \varphi(REF(q, m_o(t))) & \text{if} \quad q > t \end{cases}$$

where t is an intensity value from a given image I, REF is a restricted equivalence function [9], and $m_o(t)$ and $m_b(t)$ are the mean of the intensities of the object and background respectively. The best fuzzy set associated to the image I, Q_t, is obtained from the value of t that maximizes $SM(1, Q_t)$ [10].

$$SM(1, Q_t) = \sum_{q=0}^{L-1} h(q) REF(1, \mu_{Q_t}(q))$$

The construction of the IVFSs is done using the theory of generators [7]. The interval-valued Fuzzy Set $\widetilde{Q} \in IVFSs(X)$ is defined as follows:

$$Q_\alpha = \left\{ \left(x, \left[\mu_Q^\alpha(x), \mu_Q^{\frac{1}{\alpha}}(x) \right] \right) | x \in X \right\}$$

The parameter α represents an estimation of the uncertainty that one supposes there is in the representation of the image as a fuzzy set.

The matrix of disparities is constructed by searching the most similar window in one of the images to a window in the other image and calculating the disparity between them (along the x-axis). Since the images are represented by IVFSs, the similarity between windows is calculated with the use of Interval valued similarities [8, 43].

The Interval valued similarity used has a result an interval (instead of a value in $[0,1]$):

$$SM(x,y) = 1 - |x - y|$$

with the K_α operator [1].

$$K_\alpha = a + \alpha(b - a), \alpha = 0.5$$

Discussion

Both methodologies rely on a similar approach, showing promising results when compared to similar ones. However further work is needed to improve the matching algorithms in order to overcome problems such as occlusions, noise etc.

9 Classification

In [61], an interval type-2 fuzzy logic system that helps radiologists to detect microcalcification in mammograms is developed. There is a preprocessing of the image in which four image features are obtained: B-descriptor, D-descriptor, average intensity inside boundary, and intensity difference between inside and outside boundaries. These values are obtained for each area supposed to be a microcalcification. Then the system assigns a degree of lession to each area. The authors used an interval type 2 fuzzy system due to the uncertainty present in the calculation of B-descriptor and D-descriptor. So Interval type 2 membership functions could deal with these uncertainty in the fuzzification process.

Also John [31] have studied the use of type-2 fuzzy sets in the modeling of medical image perception. Although it's not an image processing technique the type 2 fuzzy sets were used to model the nurse opinions of tibial images. Maybe an improvement of such works could be that automatic image processing algorithms extract these characteristics from the images.

10 Final Remarks and Future Trends

As a brief resume of the conclusions of all the works reviewed using extensions of fuzzy sets in image processing, authors claim that the motivation of using extensions is their capability to deal with uncertainty present in all image processing steps. Looking to the results presented we can say that this research line is really promising. However we have found two points that we must notice:

1. All works that have been revised use interval type 2 fuzzy sets, although in the title are presented as type-2 fuzzy sets. This occurs maybe due to the complexity of dealing with real type-2 fuzzy sets or due to the fact that it's really difficult to define a type-2 fuzzy set.
2. Works dealing with intuitionistic fuzzy sets, do not use the complete information given by the intuitionistic fuzzy sets. The non-membership does not represent anything. Authors use the hesitation index π but as we have seen in the introduction it is equivalent to the length of an interval in IVFS. A really important conclusion is that all works can be done using Interval valued fuzzy sets. Therefore a problem of notation and work visibility is derived.

Hence future research, form the extensions point of view, must focus on problems that definition and computation of general type 2 fuzzy sets are tractable.

Also, finding applications or image representations in which membership and non-membership can be generated independently in order to use all of the power of A-IFSs, should be researched.

From the point of view of image processing algorithms point of view, extensions of fuzzy sets can also be used in video summarization or content image retrieval.

Acknowledgements. This work has been partially supported by Accoes integradas Luso-Espanholas 2007, POCI 2010 (Programa Operacional Ciencia e Investigacao 2010), Acciones Integradas Hispano-Portuguesas, ref. HP2006-0128 (MEC), and by the Spanish Ministry of Education and Science under project reference TIN2007-65981.

References

1. Atanassov, K.: Intuitionistic fuzzy sets. Fuzzy Sets and Systems 20, 87–96 (1986)
2. Atanassov, K.: Intuitionistic Fuzzy Sets. In: Theory and Applications. Physica, Heidelberg (1999)
3. Basu, K., Deb, R., Pattanaik, P.K.: Soft sets: an ordinal formulation of vagueness with some applications to the theory of choice. Fuzzy Sets and Systems 45, 45–58 (1992)
4. Bigand, A., Colot, O.: Fuzzy filter based on interval-valued fuzzy sets for image filtering. Fuzzy Sets and Systems (2009), doi:10.1016/j.fss.2009.03.010
5. Burillo, P., Bustince, H.: Entropy on intuitionistic fuzzy sets and on interval-valued fuzzy sets. Fuzzy Sets and Systems 78, 305–316 (1996)
6. Bustince, H., Burillo, P.: Vague sets are intuitionistic fuzzy sets. Fuzzy Sets and Systems 79, 403–405 (1996)
7. Bustince, H., Kacprzyk, J., Mohedano, V.: Intuitionistic Fuzzy generators. Application to Intuitionistic Fuzzy complementation. Fuzzy Sets and Systems 114, 485–504 (2000)
8. Bustince, H.: Indicator of inclusion grade for interval-valued fuzzy sets. Application to approximate reasoning based on interval-valued fuzzy sets. International Journal of Approximate Reasoning 23(3), 137–209 (2000)
9. Bustince, H., Barrenechea, E., Pagola, M.: Restricted Equivalence Functions. Fuzzy Sets and Systems 157, 2333–2346 (2006)
10. Bustince, H., Barrenechea, E., Pagola, M.: Image thresholding using restricted equivalence functions and maximizing the measures of similarity. Fuzzy Sets and Systems 158, 496–516 (2007)
11. Bustince, H., Pagola, M., Barrenechea, E., Orduna, R.: Representation of uncertainty associated with the fuzzification of an image by means of interval type 2 fuzzy sets. Application to threshold computing. In: Proceedings of Eurofuse Workshop: New Trends in Preference Modelling, EUROFUSE (Spain), pp. 73–78 (2007)
12. Bustince, H., Pagola, M., Melo-Pinto, P., Barrenechea, E., Couto, P.: Use of Atanassov's Intuitionistic Fuzzy Sets for modelling the uncertainty of the thresholds associated to an image. In: Fuzzy Sets and Their Extensions: Representation, Aggregation and Models. Intelligent Systems from Decision Making to Data Mining, Web Intelligence and Computer Vision. Springer, Heidelberg (2008)
13. Bustince, H., Mohedano, V., Barrenechea, E., Pagola, M.: An algorithm for calculating the threshold of an image representing uncertainty through A-IFSs. In: Proceedings of Information Processing and Management of Uncertainty in Knowledge-Based Systems, IPMU, Paris, pp. 2383–2390 (2006)

14. Bustince, H., Barrenechea, E., Pagola, M., Orduna, R.: Image Thresholding Computation Using Atanassov's Intuitionistic Fuzzy Sets. Journal of Advanced Computational Intelligence and Intelligent Informatics 11(2), 187–194 (2007)
15. Bustince, H., Barrenechea, E., Pagola, M.: Generation of interval-valued fuzzy and Atanassov's intuitionistic fuzzy connectives from fuzzy conectives and from K_α operators. Laws for conjunctions and disjunctions. Amplitude. International Journal of Inteligent systems 23, 680–714 (2008)
16. Bustince, H., Barrenechea, E., Pagola, M., Orduna, R.: Construction of interval type 2 fuzzy images to represent images in grayscale. In: False edges, Proceedings of IEEE International Conference on Fuzzy Systems, London, pp. 73–78 (2007)
17. Bustince, H., Villanueva, D., Pagola, M., Barrenechea, E., orduna, R., Fernandez, J., Olagoitia, J., Melo-Pinto, P., Couto, P.: Stereo Matching Algorithm using Interval Valued Fuzzy Similarity. In: FLINS 2008 - 8th International FLINS Conference on Computational Intelligence in Decision and Control, Spain, pp. 1099–1104 (2008)
18. Bustince, H., Barrenechea, E., Pagola, M., Fernandez, J.: Interval-valued fuzzy sets constructed from matrices: Application to edge detection. Fuzzy Sets and Systems (2009), doi:10.1016/j.fss.2008.08.005
19. Bustince, H., Pagola, M., Barrenechea, E., Fernandez, J., Melo-Pinto, P., Couto, P., Tizhoosh, H.R., Montero, J.: Ignorance functions. An application to the calculation of the threshold in prostate ultrasound images. Fuzzy Sets and Systems (2009), doi:10.1016/j.fss.2009.03.005
20. Bustince, H., Artola, G., Pagola, M., Barrenechea, E., Tizhoosh, H.: Sistema neurodifuso intervalo-valorado aplicado a la segmentacion de imagenes de ultrasonidos. In: XIV Congreso Espaol Sobre Tecnologias y Logica Fuzzy, Spain (2008)
21. Canny, J.: A computational approach to edge detection. IEEE Transactions on Pattern Analysis and Machine Intelligence 8, 679–698 (1986)
22. Chaira, T., Ray, A.K.: A new measure using intuitionistic fuzzy set theory and its application to edge detection. Applied Soft Computing 8(2), 919–927 (2008)
23. Chaira, T., Ray, A.K.: Segmentation using fuzzy divergence. Pattern Recognition Letters 24, 1837–1844 (2003)
24. Cheng, H., Jiang, X., Wang, J.: Color image segmentation based on homogram thresholding and region merging. Pattern Recognition 35(2), 373–393 (2002)
25. Deng, J.L.: Introduction to grey system theory. Journal of Grey Systems 1, 1–24 (1989)
26. Deschrijver, G., Kerre, E.E.: On the relationship between some extensions of fuzzy set theory. Fuzzy Sets and Systems 133(2), 227–235 (2003)
27. Deschrijver, G., Kerre, E.E.: On the position of intuitionistic fuzzy set theory in the framework of theories modelling imprecision. Information Sciences 177, 1860–1866 (2007)
28. Ensafi, P., Tizhoosh, H.: Type-2 fuzzy image enhancement. In: Kamel, M.S., Campilho, A.C. (eds.) ICIAR 2005. LNCS, vol. 3656, pp. 159–166. Springer, Heidelberg (2005)
29. de Haan, G., Bellers, E.B.: Deinterlacing - an overview. Proceedings of the IEEE 86(9), 1839–1857 (1998)
30. Hirota, K.: Concepts of probabilistic sets. Fuzzy Sets and Systems 5, 31–46 (1981)
31. John, R.I., Innocent, P.R., Barnes, M.R.: Neuro-fuzzy clustering of radiographic tibia image data using type 2 fuzzy sets. Information Sciences 125, 65–82 (2000)
32. Forero, M.G.: Fuzzy thresholding and histogram analysis. In: Nachtegael, M., Van der Weken, D., Van de Ville, D., Kerre, E.E. (eds.) Fuzzy Filters for Image Processing, pp. 129–152. Springer, Heidelberg (2003)
33. Gau, W.L., Buehrer, D.J.: Vague sets. IEEE Transactions on Systems, Man and Cybernetics 23(2), 751–759 (1993)

34. Grattan-Guinness, I.: Fuzzy membership mapped onto interval and many-valued quantities. Z. Math. Logik Grundlag. Mathe. 22, 149–160 (1976)
35. Gupta, M.M., Knopf, G.K., Nikiforuk, P.N.: Edge perception using fuzzy logic, in Fuzzy Computing. In: Gupta, M.M., Yamakawa, T. (eds.), pp. 35–51. Elsevier Science Publishers, Amsterdam (1988)
36. Huang, L.K., Wang, M.J.: Image thresholding by minimizing the measure of fuzziness. Pattern recognition 28(1), 41–51 (1995)
37. Jack, K.: Video Demystified a Handbook for the Digital Engineer. Elsevier, Amsterdam (2005)
38. Jeon, G., Anisetti, M., Bellandi, V., Damiani, E., Jeong, J.: Designing of a type-2 fuzzy logic filter for improving edge-preserving restoration of interlaced-to-progressive conversion. Inform. Sci. (2009), doi:10.1016/j.ins.2009.01.044
39. Jeon, G., Anisetti, M., Kim, D., Bellandi, V., Damiani, E., Jeong, J.: Fuzzy rough sets hybrid scheme for motion and scene complexity adaptive deinterlacing. Image and Vision Computing (2009), doi:10.1016/j.imavis.2008.06.001
40. Hwang, C., Rhee, F.C.-H.: Uncertain Fuzzy Clustering: Interval Type-2 Fuzzy Approach to *C*-Means. IEEE Transactions on Fuzzy Systems 15(1), 107–120 (2007)
41. Mendel, J.M., John, R.I.: Type-2 fuzzy sets made simple. IEEE Transactions on Fuzzy Systems 10(2), 117–127 (2002)
42. Mendel, J.M.: Uncertain Rule-Based Fuzzy Logic Systems. Prentice-Hall, Upper Saddle River (2001)
43. Mendel, J.M.: Advances in type-2 fuzzy sets and systems. Information Sciences 177, 84–110 (2007)
44. Mendoza, O., Melin, P., Licea, G.: Fuzzy Inference Systems Type-1 and Type-2 for Digital Images Edge Detection. Journal of Engineering Letters 15(1), 45–52 (2007)
45. Mendoza, O., Melin, P., Licea, G.: A new method for edge detection in image processing using interval type-2 fuzzy logic. In: Proceedings of Granular Computing, pp. 151–156 (2007)
46. Mushrif, M.M., Ray, A.K.: Color image segmentation: Rough-set theoretic approach. Pattern Recognition Letters 29(4), 483–493 (2008)
47. Nieradka, G.: Intuitionistic Fuzzy Sets applied to stereo matching problem. In: IWIFSGN 2007, pp. 161–171. Warsaw, Poland (2007)
48. Tehami, S., Bigand, A., Colot, O.: Color Image Segmentation Based on Type-2 Fuzzy Sets and Region Merging. In: Blanc-Talon, J., Philips, W., Popescu, D., Scheunders, P. (eds.) ACIVS 2007. LNCS, vol. 4678, pp. 943–954. Springer, Heidelberg (2007)
49. Mitchell, H.B.: Pattern recognition using type II fuzzy sets. Information Sciences 170, 409–418 (2005)
50. Montero, J., Gómez, D., Bustince, H.: On the relevance of some families of fuzzy sets. Fuzzy Sets and Systems 158, 2429–2442 (2007)
51. Otsu, N.: A threshold selection method from gray level histograms. IEEE Transactions on Systems, Man and Cybernetics 9, 62–66 (1979)
52. Pal, S.K., King, R.A., Hashim, A.A.: Automatic grey level thresholding through index of fuzziness and entropy. Pattern Recognition Letters 1(3), 141–146 (1983)
53. Russo, F.: FIRE operators for image processing. Fuzzy Sets and Systems 103, 256–275 (1999)
54. Sambuc, R.: Function Φ-Flous. In: Application a l'aide au Diagnostic en Pathologie Thyroidienne. These de Doctorat en Medicine. University of Marseille (1975)
55. Scharstein, D., Szeliski, R.: A taxonomy and evaluation of dense two-frame stereo correpondence algorithms. International Journal of Computer Vision 47, 7–42 (2002)

56. Szmidt, E., Kacprzyk, J.: Entropy and similarity of intuitionistic fuzzy sets. In: Proc. Information Processing and Management of Uncertainty in Knowledge-Based Systems, Paris, France, pp. 2375–2382 (2006)
57. Sun, Z., Meng, G.: An image filter for eliminating impulse noise based on type-2 fuzzy sets. In: ICALIP 2008. International Conference on Audio, Language and Image Processing, pp. 1278–1282 (2008)
58. Tizhoosh, H.R.: Image thresholding using type-2 fuzzy sets. Pattern Recognition 38, 2363–2372 (2005)
59. Tizhoosh, H., Krel, G., Muchaelis, B.: Locally Adaptive Fuzzy Image Enhancement. In: proceedings of 5th fuzzy days Computational Intelligence, Theory and Applications, pp. 272–276 (1997)
60. Tolt, G., Kalaykov, I.: Measured based on fuzzy similarity for stereo matching of color images. Soft Computing 10, 1117–1126 (2006)
61. Thovutikul, S., Auephanwiriyakul, S., Theera-Umpon, N.: Microcalcification Detection in Mammograms Using Interval Type-2 Fuzzy Logic System. In: Proc. FUZZIEEE, pp. 1427–1431 (2007)
62. Tulin Yildrim, M., Basturk, A., Emin Yuksel, M.: A Detail-Preserving Type-2 Fuzzy Logic Filter for Impulse Noise Removal from Digital Images. In: Proc. FUZZIEEE, U.K, pp. 751–756 (2007)
63. Tulin Yildrim, M., Basturk, A., Emin Yuksel, M.: Impulse Noise Removal From Digital Images by a Detail-Preserving Filter based on Type-2 Fuzzy Logic. IEEE Transactions on Fuzzy Systems 16(4), 751–756 (2008)
64. Vlachos, I.K., Sergiadis, G.D.: Intuitionistic fuzzy information - Applications to pattern recognition. Pattern Recognition Letters 28, 197–206 (2007)
65. Vlachos, I., Sergiadis, G.: The role of entropy in intuitionistic fuzzy contrast enhancement. In: Melin, P., Castillo, O., Aguilar, L.T., Kacprzyk, J., Pedrycz, W. (eds.) IFSA 2007. LNCS (LNAI), vol. 4529, pp. 104–113. Springer, Heidelberg (2007)
66. Wang, S.T., Chung, F.L., Hu, D.W., Wu, X.S.: A new Gaussian noise filter based on interval type-2 fuzzy logic systems. Soft Computing 9, 398–406 (2005)
67. Wei, S., Zeng-qi, S.: Research on Type-2 Fuzzy Logic System and its application. Fuzzy Systems and Mathematics 19, 126–135 (2005)
68. Emin Yuksel, M., Senior Member, IEEE, Borlu, M.: Accurate Segmentation of Dermoscopic Images by Image Thresholding Based on Type-2 Fuzzy Logic. IEEE Transactions on Fuzzy Systems (2009), doi:10.1109/TFUZZ.2009.2018300
69. Zadeh, L.A.: Fuzzy sets. Information Control 8, 338–353 (1965)
70. Zadeh, L.A.: The concept of a linguistic variable and its application to approximate reasoning – I. Information Sciences 8, 199–249 (1975)
71. Sun, Z., Meng, G.: An image filter for eliminating impulse noise based on type-2 fuzzy sets. In: International Conference on Audio, Language and Image Processing ICALIP 2008, pp. 1278–1282 (2008)

Interval Type-2 Fuzzy Cellular Model Applied to the Dynamics of a Uni-specific Population Induced by Environment Variations

Cecilia Leal-Ramirez[1], Oscar Castillo[2], and Antonio Rodriguez-Diaz[1]

[1] Facultad de Ciencias Quimicas e Ingenieria, Universidad Autonoma de Baja California, Tijuana, BC, 22390 Mexico
cleal@cicese.mx, ardiaz@uabc.mx
[2] Division de Estudios de Posgrado e Investigacion, Instituto Tecnologico de Tijuana, Tijuana, BC, 22414 Mexico
ocastillo@hasfamx.org

Abstract. As part of the research about the applicability of the Fuzzy Cellular Models (FCM) to simulate complex dynamics ecological systems, this paper presents an Interval Type-2 Fuzzy Cellular Model (IT2-FCM) applied to the dynamics of a uni-specific population. In ecology is known that in the dynamics of all population, the reproduction, mortality and emigration rates are not constants, and its variability is induced by a combination of environment factors. All kind of populations that are living together in a determinate place, and the physical factors which they interact with, compose a community. Each population o physical factor inside has its own dynamics, which also presents uncertainty given by combined effects among other environment factors inside the same community. In our model this uncertainty is represented by interval type-2 fuzzy sets, with the goal of show whether the trajectories described by the dynamics of the population present a better stability in the time and space. The validation of the model was made within a comparative frame using the results of another research where a FCM were used to describe the dynamics of a population.

1 Introduction

In nature all populations living together in a determined space and the environment physical factors with which they interact compose a community. The populations and physical factors have its own dynamics, which depend of the dynamics of many other populations or factors of the same community. Therefore, fluctuations in the size of a population can have deep effect on other species [2]. The representation of this system by means of mathematical models is a quite complicated task because in any community there are many variables involucrate and all of them are full of uncertainty.

The creation of a great variety of mathematical models on the dynamics of populations has been the main objective of the ecology in the last two hundred years [15]. The formulation of the models is focused specifically to the acquisition of new knowledge from the mathematical development, which allows to draw greater conclusions and to obtain more complete results. Nevertheless, the fundamental and

P. Melin et al. (Eds.): Bio-inspired Hybrid Intelligent Systems, SCI 256, pp. 33–47.
springerlink.com

natural question still exists on, which ecological model could be more accurate to describe dynamics of a population? [20].

In computation science a great variety of theories has been developed and applied in problems of different areas of the knowledge [3]. Fuzzy cellular model (FCM) was created as a combination from two of these theories, fuzzy logic originally introduced by [22] to model vagueness and uncertainty in the real world and automata cellular (AC) introduced by [16] to model complex dynamic systems at the time and in the space. This kind of model has been used mainly as populational growth model because it can represent the natural behavior of the ecological processes, for example in the study of the urban growth [13] and in the propagation of epidemics [6].

A FCM was constructed by [10] applied to the dynamics of population wherein the reproduction, mortality and emigration rates vary at the time and in the space according to the changes in the environment, which are produced by space and food available in each cell. The [10] model is sustained on some ecology laws emerged from population growth models established by [12], [19], [1]. Their results demonstrated that a FCM represents the population growth better than the others models.

Fuzzy logic plays an important role in the development of ecological control problems because it is possible to simulate the effects produced by environment factors in the dynamics of a population by means of the representation of its behavior throughout fuzzy rules [7]. The FCM constructed by [10] use a type-1 fuzzy logic system based in type-1 fuzzy sets for the control of these effects. However, this system cannot fully handle the uncertainties present in the model. A better way to handle such uncertainties is using a type-2 fuzzy logic system based in type-2 fuzzy sets, this system provide us with more parameters and more design degrees of freedom [18].

The type-2 fuzzy sets were originally presented by Zadeh in 1975 and are essentially "fuzzy fuzzy" sets where the fuzzy degree of membership is a type-1 fuzzy set [22], [9]. Unfortunately these sets are more difficult to handle computationally and to understand that type-1 fuzzy sets. Mendel and Liang [11], [14] characterized type-2 fuzzy sets with a superior membership function and an inferior membership function; these two functions can be represented each one by a type-1 fuzzy set membership function. The interval between these two functions represents the footprint of uncertainty (FOU), which is used to characterize a type-2 fuzzy set. Interval type-2 fuzzy sets can be used when the circumstances are too uncertain to determine exact membership grades.

The objective of this paper is to construct an interval type-2 fuzzy cellular model (IT2-FCM) applied to the dynamics of a population, taking the FCM's structure proposed by [10] and substituting its type-1 fuzzy system by an interval type-2 fuzzy system. The idea is to use interval type-2 fuzzy sets to represent the uncertainty interval produced by the dynamics of other populations or physical factors that affect the dynamics of a specific population inside a community. It is difficult to represent the environment uncertainty that induces the dynamics of a population when many factors are considered because these depend on another environment which induces its own dynamics.

In the second section of this paper the architecture of the model is presented and a description on the interval type-2 fuzzy logic system (IT2-FLS) is detailed.

In the third section an implementation is made using a simulation program. The trajectories described by the population model behaviour under different study cases are shown and compared with the obtained ones by [10]. In a last section conclusions about the representativeness and applicability of this kind of model are presented.

2 Interval Type-2 Fuzzy Cellular Model

2.1 Fuzzy Cellular Model Structure

Because natural systems contain many components whose states are dependent on numerous interactions, it is difficult to construct realistic models that match the natural level of complexity [4]. It is for this reason that simple rule-based models, such as fuzzy cellular models, can be utilized to approach large-scale problems. In this paper fuzzy cellular model proposed by [10] is modified to work as interval type-2 fuzzy cellular model (IT2-FCM), which is applied to the dynamics of a uni-specific population induced by the variation of the environment. The FCM's construction is described step-by-step in [10]. However, its final formal definition is presented and described for this study

$$N(C(i,j),t+1) = \tau_a(\cdot)N(C(i,j),t) - \tau_b(\cdot)N(C(i,j),t) - \tau_c(\cdot)N(C(i,j),t) + \prod_r(k,l,t) \tag{1}$$

This model is derived of the general idea that the change in the population size is mainly given by the difference from three related terms with the birth, the death and migration of the individuals, and these terms are proportional to the population size [12], then

$$\tau_a(\cdot)N(C(i,j),t),(births) \tag{2}$$

$$\tau_b(\cdot)N(C(i,j),t),(deaths) \tag{3}$$

$$\tau_c(\cdot)N(C(i,j),t),(emigrations) \tag{4}$$

$$\prod_r(k,l,t),(immigrations) \tag{5}$$

With the goal of having a population's behaviour closer to the reality, the space concept is integrated in the model (1) using the cellular automata theory, wherein the space, time and states are discrete. Because CA are rules, rather than equations, they allow for the direct consideration of knowledge that is not necessarily restricted to hard data and are particularly useful in consideration of complex systems [8].The CA is compound by $M \times N$ cells in a rectangular region in which the evolution of each cell $C(i,j)$, where $1 \leq i \leq M$ and $1 \leq j \leq N$, depends on its present state and the states of its immediate neighbouring cells. Therefore, the population distributed in the cellular space, at the time t, can emigrate towards the other cells according to a local rule.

The migration process happens when the population looks for optimal environment conditions under which can be developed, and even though all the conditions are not the optimal ones, the population will always try to be located into environments where each of these conditions is within tolerable limits that allow its normal development.

The population that emigrates towards the neighboring cells from a cell $C(i,j)$ is determined by a factor that is directly proportional to the population size (see equation 4), and

$$\prod_r(i,j) = \left\{ C(k,l) \left| \begin{array}{c} \max\{|k-i|,|l-j|\} \leq r \\ 1 \leq k \leq M; 1 \leq l \leq N \end{array} \right. \right\} \tag{6}$$

represents the neighborhood of the cell $C(i,j)$ in terms of a radius r, which is a positive integer number, and (k,l) is the coordinate of another cell where the magnitude of the difference between (i,k) and (l,j) does not exceed the value of r. Therefore, the population that immigrates from any neighboring cell $C(k,l)$ of the cell $C(i,j)$ to the cell $C(i,j)$, is expressed by equation (5).

The population size depends at any time on the existing balance among the factors that cause that the population reproduces dies and disperse themselves [xxxx]. Therefore, reproduction rate τ_r, mortality rate τ_m and emigration rate τ_e are defined as variables that depend on the state of two factors of the environment: food available (F) in the cell $C(i,j)$ and carry capacity $(K$ population density) that can afford the cell $C(i,j)$, both at the time t:

$$\tau_a(\cdot) = \tau_a(F(C(i,j),t), K(C(i,j),t)), \tag{7}$$

$$\tau_b(\cdot) = \tau_b(F(C(i,j),t), K(C(i,j),t)), \tag{8}$$

$$\tau_c(\cdot) = \tau_c(F(C(i,j),t), K(C(i,j),t)). \tag{9}$$

In next section it is presented step by step the development of an interval type-2 fuzzy system to represent the variable behaviour of the reproduction, mortality and emigration rates in the population grow model (1). It is believe that uncertanty of these factors could represent through interval fuzzy sets because it is given by the dynamics produced by other factors inside a same community wherein they are living and interacting. Therefore is posible that factors' behaviour have a range of posibilities to happen.

2.2 Interval Type-2 Fuzzy Logic System

Fuzzy logic is a tool to represent and to model uncertainty and imprecision in the real world [22]. In population dynamics, the fuzzy theory allows modeling many forms of biological organization in a more realistic way [17]. There are several extensions that could be considered and they include type-2 fuzzy sets.

A type-1 approach models uncertainty using a number between zero and one. Type-2 fuzzy sets represents uncertainty using a function which is itself a type-1 fuzzy number. Sometimes type-2 is referred to as fuzzy fuzzy because of this. Type-2 fuzzy sets can be used when the circumstances are too uncertain to determine exact membership grades [10].

Unfortunately, type-2 fuzzy sets are more difficult to use and to understand that type-1 fuzzy sets; hence, their use is not widespread yet. Type-2 fuzzy sets are

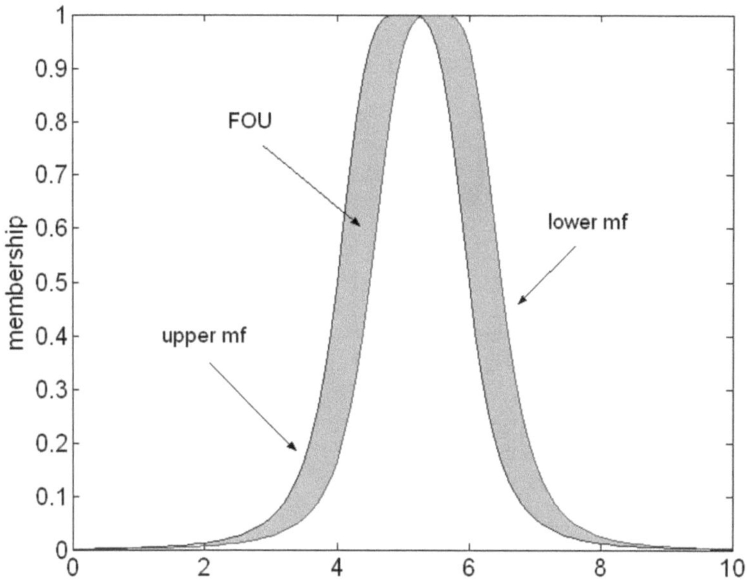

Fig. 1. FOU for Gaussian primary membership functions (mf)

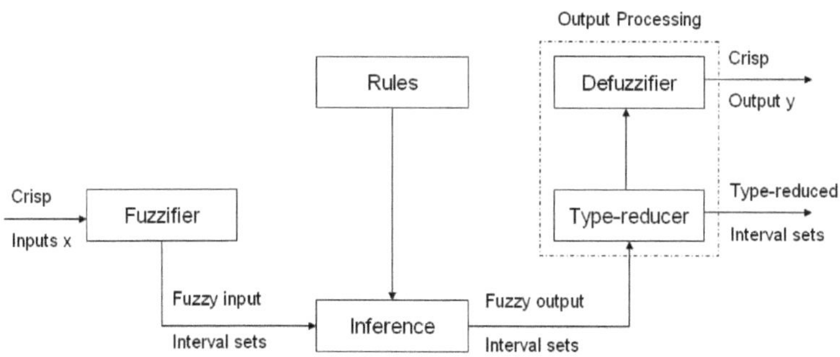

Fig. 2. Interval Type-2 Fuzzy Logic System

three-dimensional and the amplitude of their secondary membership functions (called secondary grade) can be in [0,1]. When the domain of a secondary membership function (called the primary membership) bounds a region (called the footprint of uncertainty, FOU) whose secondary grades all equal one, the resulting type-2 fuzzy set is called interval type-2 fuzzy set [18], which can easily be depicted in two dimensions instead of three (see figure 1), where the blue line denotes the upper membership function, and the green line denotes the lower membership function.

The interval type-2 fuzzy sets are not computationally as complicated as the type-2 fuzzy sets; however they can handle noisy data by making use of an interval of

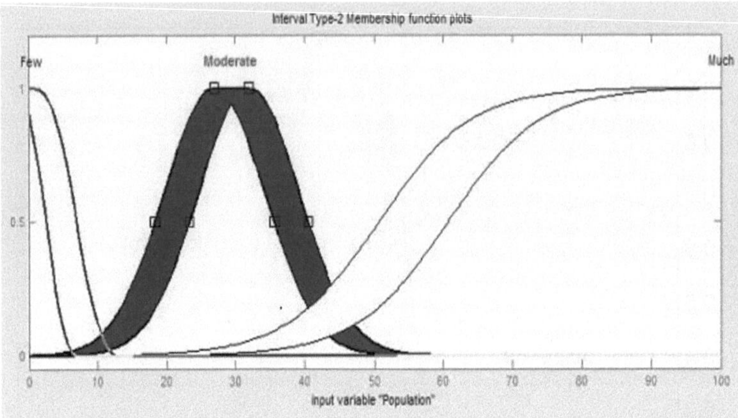

Fig. 3. Upper and Lower membership functions for the antecedent corresponding to food available in a cell at the time *t*.

uncertainty (FOU in figure 1). Systems using interval type-2 fuzzy sets are called interval type-2 fuzzy logic system (IT2-FLS), which include interval type-2 fuzzyfier, rule-base, inference engine, type-reducer and a defuzzyfier (figure 2).

All populations living together in a determined environment and the physical factors which they interact with compose a community. Populations grow or decline through the combined effects of births, deaths and migration. Therefore, the population fluctuations of a species can have deep effect on other species, including the human species [5], [2]. If the dynamics of each population depends of the effects of other species and of physical factors, then the environment of a specie is variable. That is, the exact change rate under which the population fluctuations happen is not known, therefore this rate is fuzzy. In addition, if each specie or physical factor has its own dynamics then the change rate has uncertainty in its measure. In conclusion, an uncertainty interval (FOU) can represent the uncertainty of each environment factor in model here presented.

An IT2-FLS is characterized by IF-THEN rules and its antecedent and consequent sets are now interval type-2, which are used to handle uncertainty of the environment factors in our model (figure 3).

The variable TX contains the fuzzy sets that define the reproduction rate (RR), mortality rate (MR), emigration rate (ER) and food consumption rate (CR). To represent the limitation of resource in the simulation presented in next section, the food consumption rate (CR) was included in the rules and the consequents as a state variable in each cell.

TX represents any of these rates. Nevertheless, in our system each variable has different values range in its domain and can be changed according to the behavior desired in the dynamics of a population.

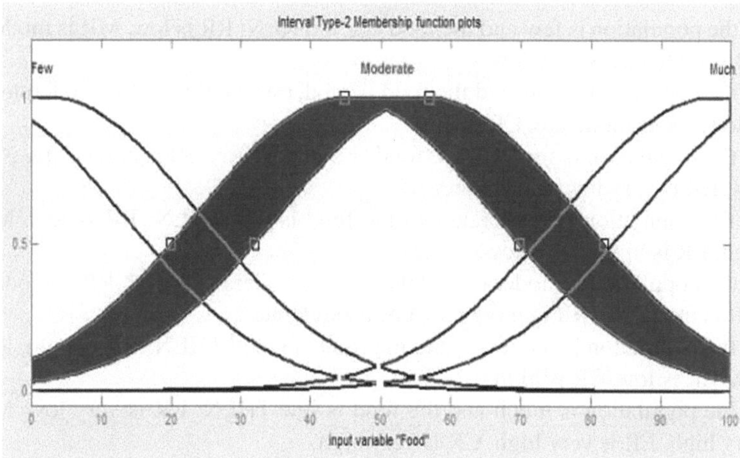

Fig. 4. Upper and Lower membership functions for the antecedent corresponding to food available in a cell at the time t

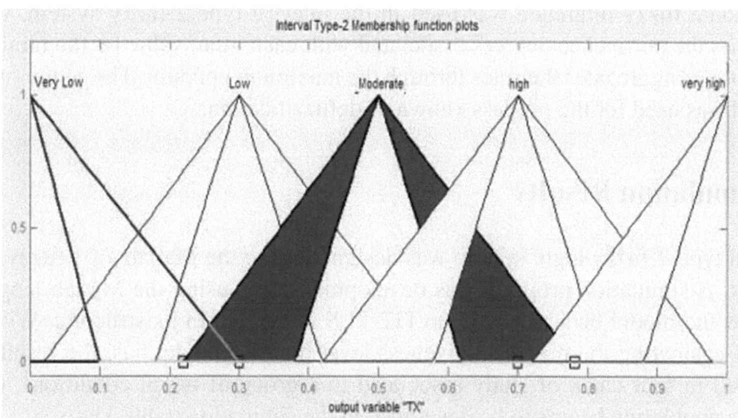

Fig. 5. Upper and Lower membership functions for each antecedent corresponding to the reproduction, mortality and emigration rates in a cell at the time t.

There is no defined pattern to determine the interval given by the upper and lower functions that charaterized the footprint of membership (FOU). A best fit is used in order to bring stability in the model and enhance surviving of the specie in a longer time interval [5] . Similar to the model proposed by [10], the conclusions established by [1], [12], [19] on the population growth and intra-specific competition are translated into the following fuzzy rules

1. IF the population is few and the food is few THEN: RR is low, MR is moderate, ER is high, CR is very low.
2. IF the population is few and the food is moderate THEN: RR is moderate, MR is low, ER is moderate, CR is low.
3. IF the population is few and the food is much THEN: RR is high, MR is very low, ER is very low, CR is low.
4. IF the population is moderate and the food is few THEN: RR is low, MR is high, ER is high, CR is low.
5. IF the population is moderate and the food is moderate THEN: RR is moderate, MR is moderate, ER is moderate, CR is moderate.
6. IF the population is moderate and the food is much THEN: RR is high, MR is low, ER is low, CR is high.
7. IF the population is much and the food is few THEN: RR is very low, MR is very high, ER is very high, CR is very high.
8. IF the population is much and the food is moderate THEN: RR is low, MR is moderate, ER is high, CR is high.
9. IF the population is much and the food is much THEN: RR is moderate, MR is moderate, ER is moderate, CR is very high.

A Mandani fuzzy inference was used in the interval type-2 fuzzy system, which combines the pertinence degrees associated with each input value by the minimum operator and aggregates the rules through the maximum operator. The center of area method was used for the process known as defuzzification.

3 Simulation Results

Interval type-2 fuzzy logic system was designed using the Matlab 7.01 fuzzy logic toolbox. A simulation program was development, also using the Matlab lenguaje, to know the model behavior with an IT2-FLS integrated in its structure. With the goal of evaluating the representativeness level that our model has, the results are classified in four cases of study associated to a group of initial conditions, under which a population begins to be developed in the simulation (table 1).

The first condition corresponding to food distribution, which can be uniformly o randomly distributed on the cellular space. In an uniform distribution each cell has the same initial amount of food, this amount is the maximum that a cell can have. Whereas in a random distribution each cell has an amount of food between 0 and its

Table 1. Initial conditions classified in four different cases of study

CASE	A	B	C	D
Food	Uniform	Uniform	Random	Random
Cells	1-5	5-10	1-5	5-10
Population	Few	Much	Few	Much

maximum. The second condition is a number between 1 and 5 or 5 and 10 selected randomly, which represents the number of cells wherein the initial population is distributed randomly, the cells are also randomly selected from cells set contained in cellular space. The last condition defines the population initial size. The simulations ran having in each iteration the parameters F and K, which are used as input to the interval type-2 fuzzy logic system.

An initial configuration is given at the beginning of a simulation run by initial population distribution on cellular space. From the ecological point of view a configuration is a possible distribution of the population in a community under which its dynamics is developed (cita). Figure 6 shows several configurations given by initial conditions grouped in case A. Whereas the configurations shows in figure 7 are given by initial conditions grouped in case B. The amount of indivuals within each cell is represented by a gray level in both figures 6 and 7. If the simulation takes initial conditions grouped in case C or D, then the configurations can be similar to ones shown in figure 6 and 7, because theses figure only show the initial distribution of the population.

If a serie of simulations are run under the same initial conditions, then a trajectories set are obtained due to random initial distribution of the population independently of any case of study (figures 8 and 9). Therefore the trajectories are a set of possible dynamics that a population can present in its development.

Figure 10 shown a comparation between the average population generated by T1-FCM and ones generated by IT2-FCM, which were produced under the same initials conditions. These populations show a logistic growth, reaching a local stability level at the different iteration time.

Using case A or B in the simulation, the population behavior begin without competition for the food because its dynamics stared with a food uniform distribution, therefore the probability of encounter among the individuals of the population grows up and a high reproduction rate is generated. The difference among case A and B is the number of cells and the population's initial size. Using case B the population reaches stability levels faster than using case A.

The initial conditions grouped in the cases C and D are similar to cases A and B respectively, the difference is the initial distribution of the food. In the cases C and D the distribution of the food is random unfavorable. Probably the population is near or more far of the food, this means that individuals have as priority its surviving and not its reproduction, therefore there is competence for the food available. That is the reason why the population generated by IT2-FCM shows lower densities than one generated by T1-FCM.

A comparation among four cases presented in figure 11, wherein figure 11-A shows as a uniform distribution favors to the population, it reaches densities in levels higher than the levels reached by population using random distribution to the food, independently of the population initial size (figures 11-C and 11-D). The random distribution of the food produces the competition among the individual for it, the individuals look at conditions more favorable by migrating. If the resource is few then they stop the reproduction and continue looking food, which imply a decrease in its density (figure 11-B). May be this case causes that the population does not

Fig. 6. Distributions of a initial population (configurations) on cellular space using case A

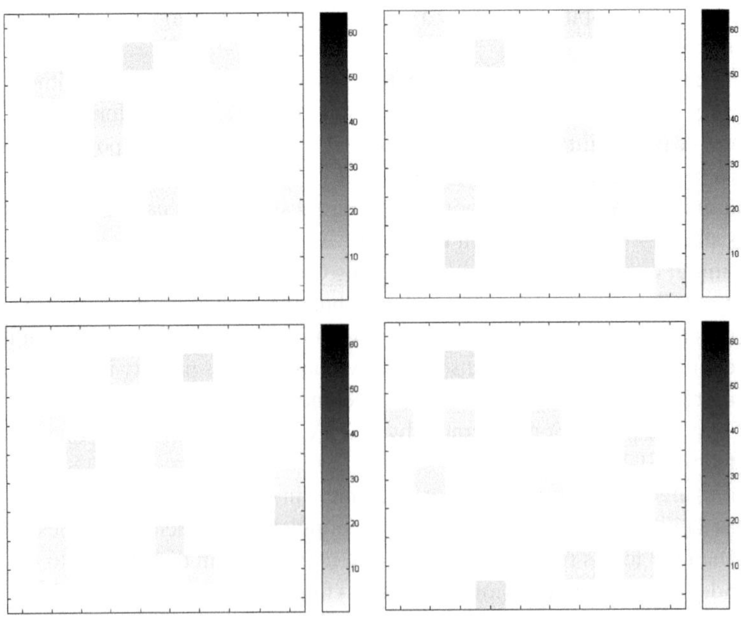

Fig. 7. Distributions of a initial population (configurations) on cellular space using case B

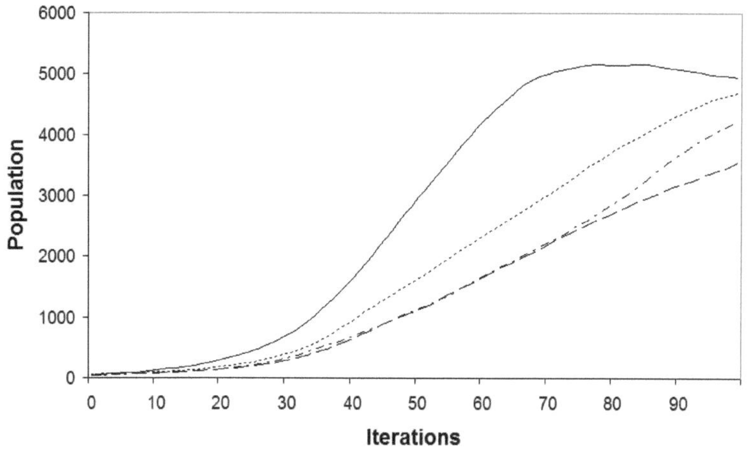

Fig. 8. Trajectories generated by configurations shown in figure 6

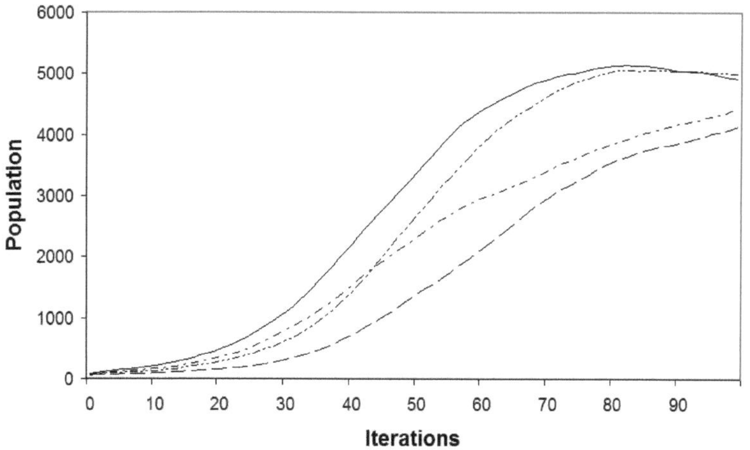

Fig. 9. Trajectories generated by configurations shown in figure 7

reaches its stability in high levels like other cases, and its densities comparated with the densities generated by TI-FCM are down of the average.

There are great difference among the levels of stability reached by both T1-FCM and IT2-FCM (figure 11). When the probability of encounter among the individuals of the population grows up fastly, the consumption of food is also increase fastly. Therefore the stability of the population is reached at times very short and its permanence is also very short because amount of available food is no sufficient to that population remains in its maximum density when its competition intra-specific begins.

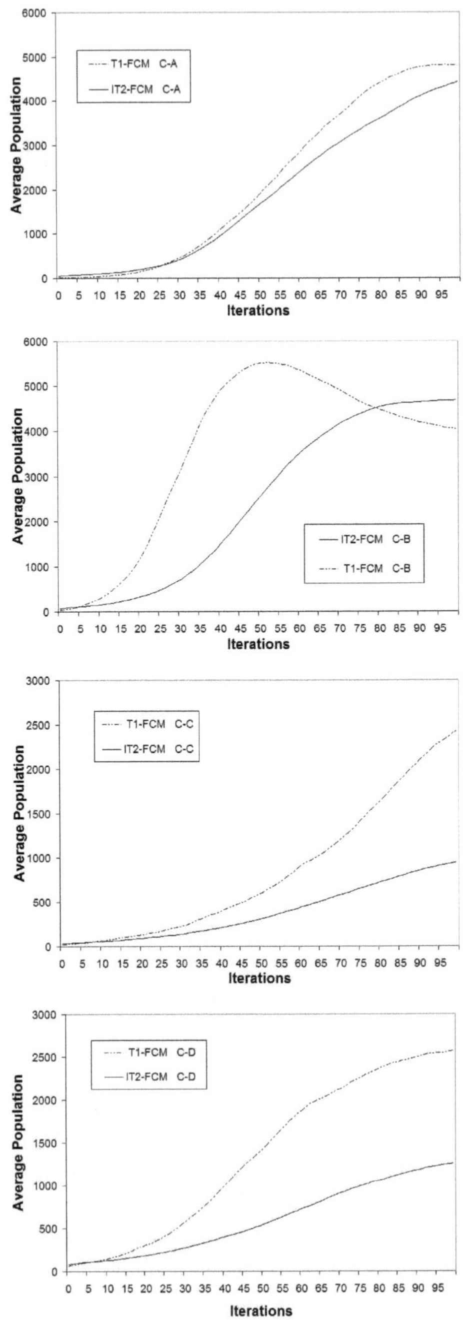

Fig. 10. Comparation among average population given by T1-FCM and IT2-FCM using initial conditions shown in table 1

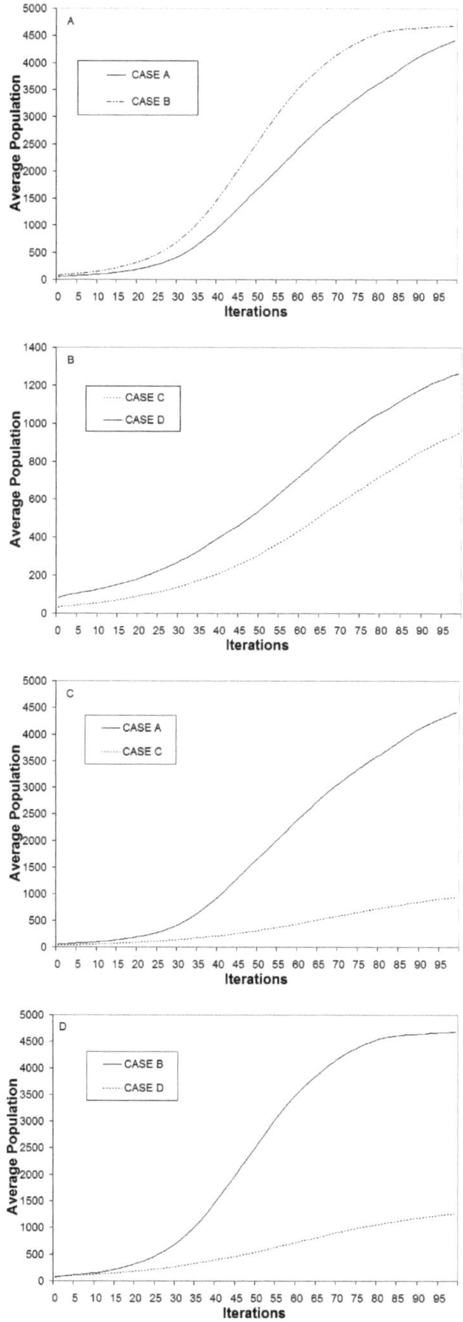

Fig. 11. Trajectories generated by IT2-FCM using the initial conditions grouped in the cases shown in table 1

4 Conclusions

Fuzzy logic mathematical methodology based in interval type-2 fuzzy sets was adopted to design interval type-2 fuzzy logic system (IT2-FLS). This system was integrated to a fuzzy cellular model applied to the population growth induced by environments factors. The simulation results were compared with the ones presented by [10] to evaluate the representativeness and applicability of this kind of model in population dynamics.

Models of population dynamics allow the valuation of the survival capacity of the species throughout the time. Cleary, the data presented in table 2 demonstrate that IT2-FCM warrants the survival of the population during a time interval longer that T1-FLS in all cases. However, in the cases C and D the IT2-FCM shows density levels lower than T1-FCM. For the biology is important to warrant the presence of the species throughout the time and density levels higher [17],[5].

Table 2. Average time intervals established by stability levels reached by population.

CASE	A	B	C	D
IT-FCM	90-95	50-55	100-105	95-100
IT2-FCM	100-120	80-120	120-150	100-130

We maintain that fuzzy-set theory represents a methodological advance for biology, and as work future will be to find the FOU appropriate for the fuzzy sets defined in this model that can represent the uncertainty of environment factors applied to a real problem that guarantees the two aspects more important for the biology inside population's dynamics study.

References

1. Alle, W.C.: Animal aggregations: a study in general sociology. University of Chicago Press, USA (1932)
2. Berryman, A.A.: Principles of population dynamics and their application. Thornes, Cheltenham (1999)
3. Fortuna, L., Rizzotto, G., Lavorgna, M., Nunnari, G., Xibili, M.G., Capponetto, R.: Soft Computing. In: New trends and Appications. Springer, London (2001)
4. Clark, J.S.: Scaling the population level: Effects of species composition in gaps and structure of a tropical rain forest. In: Ehleringer, J.R., Field, C.B. (eds.) Scaling Physiological Processes Leaf to Globe, pp. 255–285. Academic Press, New York (1993)
5. Coulson, T., Albon, S., Guinness, F.: Population substructure, local density, and calf winter survival in red deer (Cervus elaphus). Ecology 78, 852–863 (1997)
6. Di-Stefano, B., Fuks, H., Lawniczak, A.T.: Application of fuzzy logic in ca/lgca models as a way of dealing with imprecise and vague date. In: Elelctrical and Computer Engineering. Canadian Conference, vol. 1, pp. 212–217 (2000)

7. Huang, Z., Chen, S., Xia, Y.: Incorporate intelligence into an ecological system: An adaptive fuzzy control approach. Applied Mathematics and Computation 177, 243–250 (2006)
8. Jeltsch, F., Milton, S.J., Dean, W.R.J., VanRooyen, N.: Tree spacing and coexistence in semiarid savannas. J. of Ecol. 84, 583–595 (1996)
9. Karnik, N.N., Mendel, J.M.: An Introduction to Type-2 Fuzzy Logic Systems. Univ. of Southern Calif., Los Angeles (1998)
10. Leal-Ramirez, C., Castillo, O., Rodriguez-Diaz, A.: Fuzzy cellular model applied to the dynamics of a Uni-Specific Population induced by environment variations. In: Proceedings of ninth mexican international conference on computer science, pp. 211–220. IEEE Computer Society, Randall Bilof editorial (2008)
11. Liang, Q., Mendel, J.: Interval type-2 fuzzy logic systems: Theory and design. IEEE Transactions Fuzzy Systems 8, 535–550 (2000)
12. Malthus, T.R.: An Essay on the Principle of Population. J. Johnson, London (1978)
13. Mandelas, E. A., Hatzichristos, T., Prastacos, P.: In: 10th Agile international conference on geographic information science, pp. 1–9. Aalborg University, Denmark (2007)
14. Mendel, J.: Uncertain Rule-Based Fuzzy Logic Systems: Introduction and New Directions. NJ Prentice-Hall, Englewood Cliffs (2001)
15. Molina-Becerra, M.: Analisis de algunos modelos de dinamica de poblaciones estructurados por edades con y sin difusion. Phd. Universidad de Sevilla, Espana (2004)
16. Neumann, J.: Theory of self-reproducing automata. University of Illinois Press, Urbana (1966)
17. Schaefer, A.J., Wilson, C.: The fuzzy structure of populations. Can J. Zool. 80, 2235–2241 (2002)
18. Sepulveda, R., Castillo, O., Melin, P., Rodriguez-Diaz, A., Montiel, O.: Experimental study of intelligent controllers under uncertainty usineg type-1 and type-2 fuzzy logic. Information Sciences: an International Journal 177, 2023–2048 (2007)
19. Verhulst, P.F.: Notice sur la loi que la population suit dans son accrossement. Corr. Math. Phys. 10, 113–121 (1838)
20. Vladimir, I.G., Vladlena, V.G., Sean, E.M.: Population as an oscillating system. Ecological Modelling 210, 242–246 (2008)
21. Wolfram, S.: Theory and applications of cellular automata. World Scientific, Singapore (1986)
22. Zadeh, L.A.: Fuzzy set. Information and Control 8, 338–353 (1965)
23. Zadeh, L.A.: Fuzzy logic. Computer 1, 83–93 (1988)

A Genetic Programming Approach to the Design of Interest Point Operators

Gustavo Olague[1] and Leonardo Trujillo[2]

[1] Proyecto Evovisión, Departamento de Ciencias de la Computación,
División de Física Aplicada, Centro de Investigación Científica y de Educación Superior de Ensenada, Km. 107 Carretera Tijuana-Ensenada, 22860, Ensenada, BC, México
olague@cicese.mx

[2] Instituto Tecnológico de Tijuana, Av. Tecnolǵico, Fracc. Tomás Aquino, Tijuana, B.C., México
leonardo.trujillo.ttl@gmail.com

Abstract. Recently, the detection of local image feature has become an indispensable process for many image analysis or computer vision systems. In this chapter, we discuss how Genetic Programming (GP), a form of evolutionary search, can be used to automatically synthesize image operators that detect such features on digital images. The experimental results we review, confirm that artificial evolution can produce solutions that outperform many man-made designs. Moreover, we argue that GP is able to discover, and reuse, small code fragments, or building blocks, that facilitate the synthesis of image operators for point detection. Another noteworthy result is that the GP did not produce operators that rely on the auto-correlation matrix, a mathematical concept that some have considered to be the most appropriate to solve the point detection task. Hence, the GP generates operators that are conceptually simple and can still achieve a high performance on standard tests.

1 Introduction

Currently, many computer vision systems employ a local approach to feature extraction and image description. Such systems focus on small and highly invariant point features, which can be easily and efficiently detected [1, 2, 3]. The basic process was introduced in [4, 5], and proceeds as follows. Interesting image regions, each centered around a single salient point, are detected using specially designed image operators [3]. Afterwards, each region is characterized in a compact form using a specialized algorithm that takes as input the local image matrix and outputs a much smaller numerical vector that uniquely describes the region [2, 5, 6, 7]. These local regions, and their corresponding descriptive vectors, can then be used to construct discriminative [5] or generative [8] models of the objects contained within the image [9], or even of the complete scene that is being observed [10, 11]. It is expected that matches between descriptors, or more generally matches between descriptors and learnt models, will allow the system to determine when a known object or place is being observed; the basic steps of this process are shown in Figure 1.

P. Melin et al. (Eds.): Bio-inspired Hybrid Intelligent Systems, SCI 256, pp. 49–65.
springerlink.com

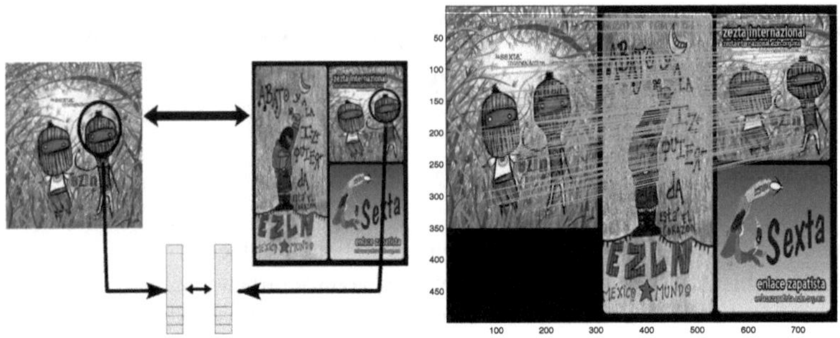

(a) Single match

(b) All of the matches

Fig. 1. The matching process between two images using local features and descriptors.

Such an approach has several advantages, including the following:

- Traditional segmentation of the image does not need to take place as a preliminary step to the higher-level recognition process.
- The approach is robust to partial occlusions, because only a small subset of local patches need to be detected in order to obtain a reliable recognition.
- Image information is sharply compressed by only focusing on a small subset of salient regions, and then by describing each patch using compact numerical vectors.

The usefulness of such an approach has been confirmed by the large number of published works that follow the basic steps outlined above. However, it is also important to understand that the performance of any such system will depend greatly on the quality of the underlying detection and description algorithms that are being used. Keeping to the former, there are perhaps dozens of proposed interest point detectors available in the computer vision literature, most of which are the direct product of a human-based approach to problem solving and/or design.

In the present chapter, we describe a genetic programming (GP) approach to the synthesis of image operators that identify salient or interesting pixels [12, 13, 14, 15]. The proposal consists on defining a set of primitive operations that are commonly found in many operators that were designed by hand, and then use GP to search for an optimal solution within the space of possible operators that these primitives define. The search process employs a fitness measure that favors operators that detect stable and dispersed image points, in order to achieve invariance to common transformations and to approximate a uniform sampling of the image plane. We are interested in describing the overall performance of the evolutionary process, showing the convergence tendencies of the algorithm, and interpreting the solutions that the GP produced using the building block hypothesis [16, 17, 18]. The experimental performance of individual operators has been thoroughly discussed in previous publications [12, 13, 19, 15], and will thus only recive minimal attention here.

The remainder of this chapter proceeds as follows. Section 2 gives a brief introduction to the interest point detection problem and reviews previously proposed solutions. Afterwards, Section 3 presents a general overview of GP, the basic algorithms is described and a review of computer vision applications is given. Then, Section 4 outlines our GP-based proposal for the synthesis of image operators and reviews the experimental results. The behavior and tendencies of the algorithm are discussed with greater detail in Section 5, by analyzing the type of building blocks that the GP constructs and employs. Finally, a summary of the chapter and concluding remarks are given in Section 6.

2 Interest Point Detection

In this section we introduce the concept of interest points and describe how these features can be detected on an image. It is important to note that interest points are part of a much larger class of local point or region-based features. Using the taxonomy given in [1, 3], we can say that interest points are detected by algorithms that primarily focus on image intensity values and only make weak assumptions regarding the underlying structure contained within the image [3].

Interest points can be described as salient image pixels that are unique and distinctive; i.e., they are quantitatively and qualitatively different from other image points, and normally interest points only represent a small fracction of the total number of image pixels [1, 20]. Furthermore, in order for these points to be useful in recognition tasks, the detection process should be invariant to common geometric or photometric transformations [1, 21, 12], and it should also be done in a simple and fast manner. If these characteristics are fulfilled then such features are considered to be *good* features to track [22].

2.1 Problem Definition

A measure of how salient or interesting each pixel is can be obtained using a mapping of the form $K(x) : \mathbb{R}^+ \rightarrow \mathbb{R}$ which we call an interest point operator. The domain of K is $\mathbb{R}^+$ because images do not contain negative intensity values. Each interest point detectors wll employ a different operator K; in this way a *detector* refers to the complete algorithmic process that extracts interest points, while an *operator* only computes the interest measure. Applying K to an image I produces what can be called an *interest image* I^*, see Figure 2. Afterwards, most detectors follow the same basic process: non-maxima suppression that eliminates pixels that are not local maxima, and a thresholding step that obtains the final set of features. Therefore, a pixel $\mathbf{x}$ is tagged as an interest point if the following conditions hold,

$$K(\mathbf{x}) > max\{K(\mathbf{x_W})|\forall \mathbf{x_W} \in \mathbf{W}, \mathbf{x_W} \neq \mathbf{x}\} \wedge K(\mathbf{x}) > h \, , \tag{1}$$

where $\mathbf{W}$ is a square neighborhood of size $n \times n$ around $\mathbf{x}$, and h is an empirically defined threshold. The first condition in Eq. (1) accounts for non-maximum suppression and the second is the thresholding step, the process is shown in Figure 2. Experiments in the current work use $n = 5$, while h depends on the operator.

Fig. 2. A look at interest point detection: left, an input image I; middle, *interest image* I^*; right, detected points after non-maximum suppression and thresholding superimposed on I. Interest points were detected with the K_{IPGP2} operator [12].

2.2 Previous Proposals and Performance Criteria

In this section only a brief overview of previous work is given, a thorough discussion is beyond the scope of this chapter. For a complete discussion and comparison of these, and other such methods, we refer the interested reader to [1, 15, 23, 3].

Over the past thirty years many detectors have been proposed, based on different interpretations of what constitutes an interest point and employing different conceptual approaches and mathematical tools in order to detect them. For example, some interest point detectors employ operators that are based on the *auto-correlation* or *second-moment matrix*. This mathematical concept captures local properties of the image gradient around each point. Hence, previous detectors have used the determinant and trace of this matrix [24, 25], as well as an eigenvalue analysis [22].

Other operators employ measures related to the local curvature around each point. For instance, [26] used the determinant of the Hessian matrix as the interest operator, and in [27] an operator is proposed that uses the gradient magnitude and the intensity with which the gradient changes direction. These operators, and others, have been shown to be mathematicaly equivalent in [28]. Other examples include operators that exploit color information, temporal information, and simple intensity variations around each point [3], and others exist.

In fact, a very large number of different proposals have been made; for instance, a few dozen are covered in the works cited above. Moreover, because of the general applicability of these methods, systems designers are presented with a problem of choice: which detector should be used to develop an image analysis system that relies on the detection of interesting point features? Therefore, reasonable and measurable criteria are required in order to make an informed decision based on the performance that each detector achieves.

When real-world applications need to be developed, probably the best approach to evaluate the performance of each detector is to use experimental criteria and a set of standard tests [1, 21]. Currently, the most common measure is the repeatability rate, which quantifies how invariant the detection process is to common types of image transformations. The repeatability rate is computed between two images taken from different viewpoints; thus, one is the base image and the other is its transformed counterpart. After detecting interest points on the base image it is possible to inspect if the same scene features are marked with interest points on the transformed image. Hence, the repeatability rate can vary from 0, meaning that no points are repeated and detection is completely unstable with respect to the transformation, to 100, which means that detection is completely stable and invariant. When the two viewpoints are related by a planar homography then computing a repeatability score is a straightforward task [1]. Other methods, however, have also been proposed, such as using mathematical axioms that describe the properties which interest points are expected to fulfill [29].

Another possible performance criterion is to consider the spread of interest points over the image plane, i.e., the amount of point dispersion over the image. Although this criterion will greatly depend upon the underlying structure of the imaged scene, many authors have stated that it is an important determining factor when choosing an interest point detector for a specific problem [20, 30, 31].

3 Genetic Programming

The evolutionary computation (EC) paradigm consists on the development of computer algorithms that base their core functionality on the basic principles of Neo-Darwinian evolution. These techniques are population-based meta-heuristics, where candidate
solutions to a particular problem are selected for reproduction, recombination and mutation, in order to explore the space of possible solutions in search of local or global optima. The selection process favors those individuals that exhibit the best performance, and the entire process is carried out iteratively, guided by the underlying structure of the fitness landscape.

Genetic programming (GP) is arguably the most advanced and complex technique used in EC, it is a generalization of the better-known, and more widely used, genetic algorithms (GAs) [32, 16, 18]. In canonical GP, each individual solution is represented through tree structures which are able to express simple computer programs, functions, or mathematical operators. Hence, it has been argued that GP provides a basic form of automatic programming, whereby one computer program evolves, or synthesizes, another. For instance, S-expressions in LISP can also be represented using trees, and GP could therefore, in theory, create any program which can also be written in the LISP programming language [16]. In general, all of the basic and advanced methods used in EC can also be included in a GP algorithm, with the main particularities being those related to the manner in which trees are created, combined and randomly modified. In practice, when applying GP to a specific problem the most important task is to define the set F of functions that can be used as tree nodes, and the set T of terminal elements that can appear as tree leaves. These sets define the search space where the GP can generate and test individual programs.

3.1 Computer Vision Applications

GP has proven to be a powerful tool in many scientific domains, achieving results which can be regarded as human-competitive in such domains as antenna design, game playing strategies, and computer vision [33, 34]. For instance, computer vision presents a large variety of problems where the goal is to allow computers and robots to interact with their environment using visual information. The difficulty of such problems is unquestionable, and consequently many researchers have turned to machine learning and EC techniques in their attempts to solve them [35, 36, 37]. GP has recently received interest in this domain because of its ability to synthesize specialized operations that can detect image features, or operators that are able to construct new features which can then be used in higher-level tasks. In a recent special issue of *Evolutionary Computation* on *Computer Vision* three of the five published papers employ GP as their main search and optimization process [37]. In what follows, we review some recent examples of GP-based applications in computer vision and image analysis.

It is possible to identify three types of GP-based approaches [38]: (1) those that employ GP to detect low-level features which have been predefined by human experts, such as corners or edges [39, 40, 12, 13, 15, 41]; (2) those that construct novel low-level features which are specific to a particular problem domain, and need not be interpretable to a human expert [38, 42, 43]; and (3) those approaches that use GP to directly solve a high-level recognition problem [44, 45, 46].

For instance, keeping to the latter two groups, works have addressed the problem of object detection [45], image classification [38], texture segmentation [46], and the analysis of synthetic aperture images [44, 42, 43], to name but a few examples. In all of these works, evolution is left unconstrained with regards to the type of information that individual solutions will extract from the image. Therefore, evolution in only guided by the performance that each individual achieves on the higher-level tasks, thus exploiting the powerful search capability that artificial evolution offers.

However, such approaches can be affected by two shortcommings which are common in EC: (a) the operations of some individual solutions can be quite unintuitive and in many cases difficut to understand; and (b) a semantic interpretation of the features that the GP generates is normally very difficult to obtain, or sometimes not even possible. On the other hand, the works from the first group, those that attempt to detect features defined by human experts, by definition will not be hampered by (b), however (a) can still be a limitation. For instance, the most common types of features in vision literature are salient points and line segments, and both have been studied using GP approaches, examples include [39, 12, 13, 15] for the former, and [40, 41] for the latter. However, in some cases it is not a trivial task to understand the logic behind some of the solutions that the GP produces [39, 40], and this reduces the confidence that researchers in other fields might have in such methods. Hence, some researchers have stressed the importance of uderstanding what the GP solutions are actually doing, in this domain [12, 13, 15, 41] and in other areas of research [47].

In the current chapter, we continue the analysis of the work done in [12, 13, 15] where interest point operators are generated using a GP algorithm. The goal is to give a more complete interpretation of the types of operations that the GP produces and thus help reduce the problems related with the shortcomings of type (a) discussed above. Here, it will be seen that the GP clearly prefers certain types of compact sub-operations, or building-blocks, which it constructs, and reconstructs in different experimental runs.

4 Synthesis of Interest Point Operators

In this section, we discuss how the design of an interest point operator is carried out automatically using a GP algorithm. First, the search space is defined through a set of primitive operations that are commonly used in many man-made designs. Then, the fitness function is introduced, which promotes the creation of operators that achieve invariant detection with a high amount of point dispersion. Finally, we briefly discuss the overall performance of the algorithm with some examples.

4.1 Primitive Operations and Fitness Evaluation

Functions and Terminals

In order to define an appropriate search space, the function and terminal sets contain simple operations that are common to many of the previously proposed interest point operators; therefore, these sets are defined as

$$\begin{aligned} F = \{&+, |+|, -, |-|, |I_{out}|, *, \div, I_{out}^2, \sqrt{I_{out}}, log_2(I_{out}), EQ(I_{out}), k \cdot I_{out}\} \\ &\cup \left\{\tfrac{\delta}{\delta x} G_{\sigma_D}, \tfrac{\delta}{\delta y} G_{\sigma_D}, G_{\sigma=1}, G_{\sigma=2}\right\}, \\ T = \{&I, L_x, L_{xx}, L_{xy}, L_{yy}, L_y\}, \end{aligned} \tag{2}$$

where I is the input image, and I_{out} can be any of the terminals in T, as well as the output of any of the functions in F; $EQ(I)$ is an histogram normalization operation; L_u is a Gaussian image derivatives along direction u; G_σ are Gaussian smoothing filters; $\frac{\delta}{\delta u} G_{\sigma_D}$ represents the derivative of a Gaussian function[1]; and a scale factor of $k = 0.05$.

Fitness Evaluation

As stated above, it is expected that fitness should promote the emergence of interest operators that perform invariat detection of highly repeatable points, and that the set of detected points should be highly dispersed over the image plane. In this work, both objectives are combined in a multiplicative manner, as follows

$$f(K) = r_{K,J}(\varepsilon) \cdot \phi_x^\alpha \cdot \phi_y^\beta \cdot N_\%^\gamma, \tag{3}$$

where $r_{K,J}(\varepsilon)$ represents the average repeatability rate of an individual operator K computed from a set J of pregresively transformed images with a localization error ε, and the terms ϕ_u promote a high point dispersion. The final term , $N_\%$ is a penalizing factor that reduces the fitness value for detectors that return less than the number of requested point. The terms ϕ_u behave like sigmoidal functions within a specified interval,

$$\phi_u = \begin{cases} \dfrac{1}{1 + e^{-a(H_u - c)}}, & when \quad H_u < H_u^{max}, \\ 0 & otherwise, \end{cases} \tag{4}$$

[1] All Gaussian filters are applied by convolution.

where H_u is the entropy value of the spatial distribution of detected interest points along direction u, on the reference image I_1 of the training set J, given by

$$H_u = -\sum P_j(u) log_2 [P_j(u)] \;, \tag{5}$$

with $P_j(\cdot)$ approximated by the histogram of interest point localizations. What is important to remark about the fitness function is that it primarily promotes a high repeatability score, and penalizes operators that obtain entropy values for point dispersion that lie outside a specified bound. We refer the interested reader to [15] where a more detailed explanation of the fitness function is provided.

4.2 Experimental Results

This section will focus on a total of 32 different executions of the GP algorithm, from which 17 useful interest operators were obtained. The reason for the discrepancy, between the number of runs and the number of operators, is the fact that some runs produced the same operator (3 times), and in others the operator that was found did not achieve a stable performance on the images that were not included during the training phase (12 times).

Run-Time Parameters

The parameters of the GP algorithm provided in Table 1 were used during all runs.

The first four parameters were set empirically with canonical values. The next three help to control code bloat. *Tree depth* is dynamically set using two maximum tree depths that limit the size of any given individual within the population. The *dynamic max depth* is a maximum tree depth that may not be surpassed by any individual unless its fitness matches or surpasses the fitness of the best individual found so far. When this happens, the *dynamic max depth* is augmented to the tree depth of the new fittest individual. Conversely, it is reduced if the new best individual has a

Table 1. General parameter settings for our GP framework.

Parameters	Description and values
Population size	50 individuals.
Generations	50.
Initialization	Ramped Half-and-Half.
Crossover	Standard crossover.
Crossover & Mutation prob.	Crossover prob. $p_c = 0.85$; mutation prob. $p_\mu = 0.15$.
Tree depth	Dynamic depth selection.
Dynamic max depth	5 levels.
Real max depth	7 levels.
Selection	Tournament selection with lexicographic parsimony pressure.
Survival	Elitism.
Fitness function parameters	$a_x = 7$, $c_x = 5.05$, $a_y = 6$, $c_y = 4.3$, $\alpha = 20$, $\beta = 20$, $\gamma = 2$.

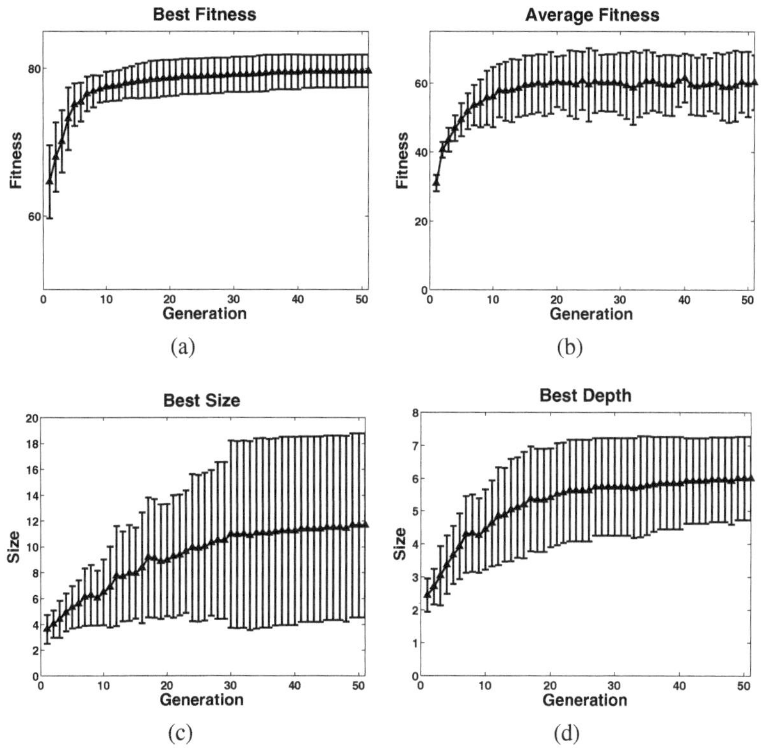

Fig. 3. Convergence plots of the GP algorithm, showing mean value from the 32 runs, and standard deviation across 50 generations: (a) Best fitness; (b) Average population fitness; (c) Size of best solution; and (d) Depth of best solution.

lower tree depth. The *real max depth* parameter is a hard limit that no individual may surpass under any circumstance. Finally, selection is carried out using a tournament with lexicographic parsimony pressure.

Convergence of the GP Algorithm

Here, we present the convergence plots of the GP algorithm, by averaging over all 32 runs and computing the standard deviation at each generation, the results are plotted in Figure 3. First, Figure 3(a) shows how the fitness of the best individual in the population changes during the run. It can be seen that fitness largely improves during the initial generations, and only slightly at the end of most runs, a characteristic which is typical to evolutionary algorithms. Nevertheless, there is a clear pattern of progressive improvements and overall optimization of the solutions. This is also shown in Figure 3(b) with the average population fitness, showing a pattern similar to that of the best fitness, but with a higher variance. Figures 3(c)-(d) present the evolution dynamics of the size of the best solution found; these figures plot, respectively, the number of nodes and the depth of the corresponding program tree.

Table 2. The best interest operators evolved by the GP algorithm described above. Table contains the following 5 columns: Name (the identification name for each operator, consistent with [15]); Operator (the mathematical expression); Run (the numbers of times this operator was generated); Fitness (the fitness of the operator); and r_J (the average repeatability rate computed on the training sequence).

Name	Operator	Runs	Fitness	$r_J \star$
$IPGP1$	$G_2 * (G_1 * I - I)$	1	79.76	94.465
$IPGP1^*$	$G_2 * \lvert I - G_2 * I \rvert^2$	2	77.91	94.78
$IPGP2$	$G_1 * [L_{xx} \cdot L_{yy} - L_{xy}^2]$	1	67.3	93.02
$IPGP3$	$G_1 * G_1 * G_1 * G_2 * G_2 * (G_1 * I - I)$	1	83.98	95.9
$IPGP4$	$G_2 * G_2 * G_2 * (G_2 * I - I)$	1	85.98	96.35
$IPGP5$	$G_1 * G_2 * \lvert I - G_1 * I \rvert^2$	1	83.86	93.22
$IPGP6$	$G_2 * G_2 * G_1 * \left(\frac{I}{G_2 * I} \right)$	1	78.13	94.84
$IPGP7$	$G_2 * \lvert 2 \cdot L_{yy} + 2 \cdot L_{xx} \rvert$	2	78.33	94.92
$IPGP8$	$G_2 * \lvert L_{xx} + 2 \cdot G_2 (L_{xx} + L_{yy})^2 \rvert$	1	73.86	90.54
$IPGP9$	$G_2 * G_2 * \lvert 2 \cdot L_{yy} + 2 \cdot L_{xx} + L_{xy} \rvert$	1	80.3	93.44
$IPGP10$	$G_2 * \lvert L_{yy} + L_{xx} \rvert$	2	77	92.81
$IPGP11$	$G_1 * \left(\frac{G_1 * I}{(G_1 * G_1 * I)^3} \right)$	1	78.23	92.44
$IPGP12$	$\frac{G_2 * I^{\frac{3}{2}}}{(G_1 * I)^{\frac{9}{4}}}$	1	72.67	91.91
$IPGP13$	$G_2 * G_2 * [(G_2 * I)(G_2 * G_1 * I - I)]$	1	85.72	96.37
$IPGP14$	$G_2 * G_2 * [(G_2 * G_2 * G_2 * I)(G_2 * G_2 * I - I)]$	1	85.94	96.5
$IPGP15$	$\frac{G_2 * [G_2 * G_2 * \lvert I - G_1 * I - G_2 * G_2 * I \rvert]}{G_2 * G_2 * I}$	1	85.81	95.15
$IPGP16$	$\frac{G_2 * G_2 * [G_1 * G_2 * I^2 - I^2]}{G_2 * G_2 * G_1 * I}$	1	84.63	96.8

For the former, Figure 3(c) shows that even if the size of the tree tends to increase with each generation, the variance is quite large. Hence, it can be argued that code bloat was effectively limited in some runs. This is also evident in Figure 3(d), where the depth of the best tree does not tend to the maximum allowed depth of 7 levels, the mean value asymptotically reaches a maximum depth of 6.

Interest point operators generated by the GP search

The best operators found by the GP were presented with great detail in [15]; therefore, only a brief description of these operators is given next. Table 2 gives a list of the best operators generated by the GP, showing their assigned name [2], the mathematical expression for each operator, the number of runs in which the operator was produced, the respective fitness, and the average repeatability rate computed on the training sequence. For clarity, in some cases the mathematical expression was sim-

[2] The acronym IPGP stands for *Interest Point* detection with *Genetic Programming*.

plified, and therefore the true individual is sometimes larger or more complex than what the expressions in Table 2 might suggest.

In [12, 13, 19, 15], the performance with respect to repeatability of the evolved operators was verified on a series of image sequences that contain different types of scenes and different transformations. It was shown that evolved operators achieved better performance than most of the man-made detectors, using original and improved codes, included in the comparison. In fact, only one, the improved version of the Harris & Stephens detector [24, 1], achieved comparable repeatability rates. Hence, the proposed GP algorithm is considered to be human-competitive for the problem of interest point detection, making it one of only 60 or so results that have achieved this recognition from the EC community [34] (p. 120).

5 Discussion

The experimental results obtained by the GP algorithm have been encouraging, and suggest that evolution is indeed capable of synthesizing useful and competitive solutions. However, the structure and functionality of some of the solutions were indeed unexpected. For instance, an early assumption was that the GP would have used elements from the auto-correlation matrix as part of the solutions it produced because it has been argued that this mathematical construct is the best tool to solve this task [29]. The auto-correlation matrix describes the gradient around each point using first order image derivatives, but these elements rarely appeared in the best solutions found, see Figure 4. Therefore, it is interesting to note that the GP algorithm did not produce operators similar to those proposed in [24, 25]. Furthermore, other terminals and functions are likewise unused ($\frac{\delta}{\delta u}G_{\sigma_D}, |+|, EQ$) or have a very low

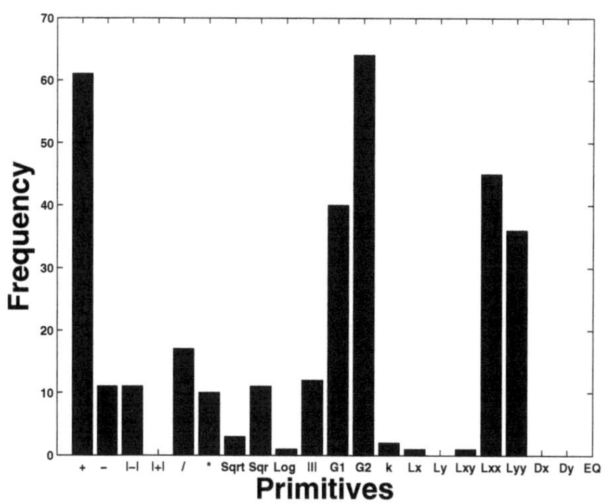

Fig. 4. Frequency with which each function or terminal from $\{F \cup T\}$ appear in the genotype of the best individuals generated by the GP.

frequency ($I^2, log, k \cdot I, L_{xy}$). It is clear that the GP is eliminating primitive functions which it finds to be of limited or no use for the point detection problem.

5.1 Building Blocks for Interest Point Operators

In what follows, we discuss the type of operators that the GP produced using arguments based on the *Building Block Hypothesis* (BBH) and the schema theorem [48, 18], two tightly linked concepts that attempt to explain how, and why, a genetic search process works.

Succinctly, the BBH states that a GA, or a GP, combines short, low-order schemata of above average fitness, or building blocks, in order to create longer and higher-order schemata (larger blocks), and that this process is progressively and iteratively carried out until a local, or hopefully global, optimum is found.

The BBH has received strong criticisms [49], and some have argued that the schema theorem is of very limited use if one wants to understand the complex dynamics of GAs [50] or GP algorithms [17]. However, the frequently stated shorcommings of the original schema theorem are not applicable to the more recent exact formulations, and therefore

> ... the [BBH] might still be considered a reasonable first approximation of the behaviour of GP ... [18](p. 108).

In this work, the building blocks we are going to discuss are sub-operations similar to what Koza originally identified as GP schemata [16], an idea that was later formalized in [17]. The main difference is that we are not particularly interested in the genotypic sub-tree within each operator, as much as we are concerned with the phenotypic, or functional, sub-operation that is being carried out. However, the following discussion does not intended to give a rigorous account of schema propagation, it is only a reasonable description of the structure and functionality that the GP solutions exhibit. Nevertheless, even an informal account can be quite revealing with regards to the manner in which the interest point detection problem could be solved.

For example, we could consider single primitives from $\{F \cup T\}$ as our most elementary candidates for possible building blocks. In this case, we can easily eliminate some of them because they appear only rarely within the genotype of the best solutions found, or sometimes not at all; for instance, first order derivatives are almost completley unused (see Figure 4). From this it follows that combining elements from the auto-correlation matrix does not seem to produce sub-trees that the GP can then use to build more complex operators. Therefore, even do operators that are based on the auto-correlation matrix [24, 25] are indeed located near local optima within the proposed fitness landscape, building such expressions with GP appears to be very difficult; i.e, the auto-correlation matrix does not provide the appropriate low order building blocks that the GP can then use to progressively evolve interest operators that achieve a better fitness measure. It could be argued that this conclusion is only applicable to the experimental runs carried out here because of the small population size and/or limited number of generations used. However, other experiments with

Table 3. Building blocks for the interest point detection problem. Columns contain the name, the mathematical operation, the number of operators that contain at least one instance of the sub-operation, and a brief description.

Name	Building Block	No.	Description
Gaussian based operation	$G_\sigma * I \# G_\gamma * I$	14	In this operation # represents any function of arity-2, and the smoothing filters can be composed by more than a single Gaussian function, with $\sigma \neq \gamma$.
Gaussian Smoothing	$G_\Sigma * I_out$	10	These operations apply a large Gaussian filter to the input image or to the output of a function node. In this case, the smoothing applied by G_Σ is $\geq$ than applying two filters with $\sigma = 2$.
Laplacian	$L_{xx} + Lyy$	15	The operation is straightforward and widely known as the Laplacian-of-Gaussian filter (LoG).

more generation and an explicit diversity preservation mechanism, also generated operators in which elements from the auto-correlation matrix were overwhelmingly absent [14]. Hence, contrary to what is suggested in [29] other locally optimal solutions for the interest point detection problem do in fact exist, and these do appear to be amenable to GP synthesis while also containing interpretable building blocks.

Table 3 presents three building blocks that recurrently appear within the genotype of the best individuals found in the 32 GP runs. Obviously, even in such a small set of individual solutions the total number of possible sub-operations can still be extremely large. However, we only list those operations that fulfill the following conditions: (1) they appear within a large proportion of the final solutions found; and (2) the hyperplane on which the schema resides contains image operations which are functionally interpretable. Table 3 presents the three operations that we consider to be functional building blocks for the point detection problem. The table shows the name, the corresponding mathematical expression, the number of operators that instantiate a version of the operation, and a brief description of each.

The first building block is a Gaussian based operation, which appears in 14 of the evolved operators. In this case, each instance contains a function of arity-2 in the place of symbol #. In eight of those operators, the operation is a difference or absolute difference, in which case it corresponds with a Difference-of-Gaussian filter (DoG). This operation is common in image processing tasks because it enhances image borders, and it has also been used for blob detection. On the other hand, on five other occasions # is instantiated with a protected division. In this case, if $\sigma \rightarrow 0$ and $\gamma > 0$, for example, then the operation gives the ration between each pixel and the Gaussian weighted mean around a local neighborhood. Such an operator can detect points which are brighter, or darker if $\sigma > 0$ and $\gamma \rightarrow 0$, than other points around it. Finally, in only a single case # was instantiated with a $+$ as part of the larger operation.

The next building block is an operation that applies a large Gaussian smoothing. In $\{F \cup T\}$, only two Gaussian filters were included, $G_{\sigma=1}$ and $G_{\sigma=2}$. However, it is evident that the GP clearly favors a larger smoothing, and to achieve this several

Gaussian functions are concatenated, sometimes as many as three or four. This type of operation was instantiated a total of ten times, when at least two or more filters with $\sigma = 2$ were used sequentially, or when more than three filters with any σ were used. A reasonable explanation for this building block can be established by considering the linear scale space theory [51], which states that scale transformations can be modeled by using Gaussian filters. Therefore, the GP is attempting to extract point features at coarser scales of observation, which is a good approach if stable and invariant detection is desired.

The third and final building block operation that was identified was the Laplacian, which appeared in fifteen different operators. The Laplacian filter is functionally similar to the DoG filter, with the latter usually used as an approximation of the former. Therefore, it is also used as a simple border detector and as a blob or scale invariant region detector [51]. Moreover, if we consider that none of the operators contained both a DoG filter and a Laplacian operation, then 23 of the 32 operators converged to operations that are functionally very similar.

6 Summary and Conclusions

In this chapter, we presented a GP-based approach to evolve image operators that detect interest points. The terminal and function sets contain primitive operations that are common in many previously proposed man-made designs. Fitness measures the stability of each operator through the repeatability rate, and promotes a uniform dispersion of detected points across the image plane. The performance of the evolved operators has been confirmed experimentally using training and testing sequences of progressively transformed images [12, 13, 19, 15].

The average over 32 different runs of the algorithm shows that fitness, of the best solution and of the population mean, tends to converge and achieve asymptotic behavior. Likewise, bloat control appears to be effective because the GP does not necessarily converge towards the maximum allowed depth; therefore, a high fitness is achieved by relatively simple operator trees.

Previously, it has been argued that the use of the auto-correlation matrix could be the best mathematical tool for solving the point detection problem [29], and several experimental results appeared to confirm such claims [1, 21], particularly for the Harris & Stephens detector [24]. However, the GP search was clearly able to find other local maxima within the space of possible operators. Moreover, the auto-correlation matrix appears to be situated on a very isolated peek within the fitness landscape, and the GP is unable to effectively create building blocks based on its components.

Nevertheless, the GP algorithm does find several sub-trees, or sub-operators, that promote a better than average performance in operators that contain them. These can be considered to be building blocks for the posed evolutionary problem, that sometimes appear as the primary component of an individual operator, or as part of a more complex operation. Therefore, it may be beneficial to use such building blocks as part of more advanced sets of functions and terminals for a future implementation of the proposed GP approach.

References

1. Schmid, C., Mohr, R., Bauckhage, C.: Evaluation of interest point detectors. International Journal of Computer Vision 37(2), 151–172 (2000)
2. Mikolajczyk, K., Schmid, C.: A performance evaluation of local descriptors. IEEE Transactions on Pattern Analysis and Machine Intelligence 27(10), 1615–1630 (2005)
3. Tuytelaars, T., Mikolajczyk, K.: Local invariant feature detectors: a survey. Found. Trends Comput. Graph. Vis. 3(3), 177–280 (2008)
4. Schmid, C., Mohr, R.: Local grayvalue invariants for image retrieval. IEEE Transactions on Pattern Analysis and Machine Intelligence 19(5), 530–534 (1997)
5. Lowe, D.G.: Object recognition from local scale-invariant features. In: Proceedings of the International Conference on Computer Vision (ICCV), Kerkyra, Corfu, Greece, September 20-25, vol. 2, pp. 1150–1157. IEEE Computer Society, Los Alamitos (1999)
6. Trujillo, L., Olague, G., Legrand, P., Lutton, E.: Regularity based descriptor computed from local image oscillations. Optics Express 15, 6140–6145 (2007)
7. Perez, C.B., Olague, G.: Learning invariant region descriptor operators with genetic programming and the f-measure. In: Proceedings of the 2008 International Conference on Pattern Recognition (ICPR 2008), pp. 1–4 (2008)
8. Bouchard, G., Triggs, B.: Hierarchical part-based visual object categorization. In: Proceedings of the 2005 IEEE Computer Society Conference on Computer Vision and Pattern Recognition (CVPR), San Diego, CA, June 20-26, 2005, vol. 1, pp. 710–715. IEEE Computer Society Press, Los Alamitos (2005)
9. Ferrari, V., Tuytelaars, T., Van Gool, L.: Simultaneous object recognition and segmentation from single or multiple model views. Int. J. Comput. Vision 67(2), 159–188 (2006)
10. Wang, J., Zha, H., Cipolla, R.: Coarse-to-fine vision-based localization by indexing scale-invariant features. IEEE Transactions on Systems, Man, and Cybernetics, Part B 36(2), 413–422 (2006)
11. Trujillo, L., Olague, G., de Vega, F.F., Lutton, E.: Evolutionary feature selection for probabilistic object recognition, novel object detection and object saliency estimation using gmms. In: Proceedings from the 18th British Machine Vision Conference, September 10-13, Warwick, UK, pp. 630–639. British Machine Vision Association (2007)
12. Trujillo, L., Olague, G.: Synthesis of interest point detectors through genetic programming. In: Cattolico, M. (ed.) Proceedings of the Genetic and Evolutionary Computation Conference (GECCO), Seattle, Washington, July 8-12, vol. 1, pp. 887–894. ACM Press, New York (2006)
13. Trujillo, L., Olague, G.: Using evolution to learn how to perform interest point detection. In: Proceedings of the 18th International Conference on Pattern Recognition (ICPR), Hong Kong, China, August 20-24, vol. 1, pp. 211–214. IEEE Computer Society Press, Los Alamitos (2006)
14. Trujillo, L., Olague, G., Lutton, E., de Vega, F.F.: Multiobjective design of operators that detect points of interest in images. In: Cattolico, M. (ed.) Proceedings of the Genetic and Evolutionary Computation Conference (GECCO), Atlanta, GA, USA, July 12-16, pp. 1299–1306. ACM, New York (2008)
15. Trujillo, L., Olague, G.: Automated design of image operators that detect interest points. Evolutionary Computation 16(4), 483–507 (2008)
16. Koza, J.R.: Genetic Programming: On the Programming of Computers by Means of Natural Selection. MIT Press, Cambridge (1992)

17. O'Reilly, U.-M., Oppacher, F.: The troubling aspects of a building block hypothesis for genetic programming. In: Proceedings of the Third Workshop on Foundations of Genetic Algorithms (FOGA), Estes Park, Colorado, USA, FOGA, July 31 - August 2, 1994, pp. 73–88. Morgan Kaufmann, San Francisco (1994)
18. Langdon, W.B., Poli, R.: Foundations of Genetic Programming. Springer, New York (2002)
19. Trujillo, L., Olague, G.: Scale invariance for evolved interest operators. In: Giacobini, M. (ed.) EvoWorkshops 2007. LNCS, vol. 4448, pp. 423–430. Springer, Heidelberg (2007)
20. McGlone, J., et al. (eds.): Manual of photogrammetry. American Society of Photogrammetry and Remote Sensing (2004)
21. Tissainayagam, P., Suter, D.: Assessing the performance of corner detectors for point feature tracking applications. Image Vision Comput. 22(8), 663–679 (2004)
22. Shi, J., Tomasi, C.: Good features to track. In: Proceedings of the 1994 IEEE Conference on Computer Vision and Pattern Recognition (CVPR 1994), Seattle, WA, USA, June 1994, pp. 593–600. IEEE Computer Society, Los Alamitos (1994)
23. Olague, G., Hernández, B.: A new accurate and flexible model based multi-corner detector for measurement and recognition. Pattern Recognition Letters 26(1), 27–41 (2005)
24. Harris, C., Stephens, M.: A combined corner and edge detector. In: Proceedings from the Fourth Alvey Vision Conference, vol. 15, pp. 147–151 (1988)
25. Förstner, W., Gülch, E.: A fast operator for detection and precise location of distinct points, corners and centres of circular features. In: ISPRS Intercommission Conference on fast processing of photogrammetric data, pp. 149–155 (1987)
26. Beaudet, P.R.: Rotational invariant image operators. In: Proceedings of the 4th International Joint Conference on Pattern Recognition (ICPR), Tokyo, Japan, pp. 579–583 (1978)
27. Kitchen, L., Rosenfeld, A.: Gray-level corner detection. Pattern Recognition Letters 1, 95–102 (1982)
28. Noble, A.: Descriptions of Image Surfaces. PhD thesis, Department of Engineering Science. Oxford University, Oxford (1989)
29. Kenney, C.S., Zuliani, M., Manjunath, B.S.: An axiomatic approach to corner detection. In: International Conference on Computer Vision and Pattern Recognition (CVPR) (June 2005)
30. Davison, A.J., Molton, N.D.: Monoslam: Real-time single camera slam. IEEE Trans. Pattern Anal. Mach. Intell. 29(6), 1052–1067 (2007); Member-Ian D. Reid and Member-Olivier Stasse
31. Yang, G., Stewart, C.V., Sofka, M., Tsai, C.-L.: Registration of challenging image pairs: Initialization, estimation, and decision. IEEE Trans. Pattern Anal. Mach. Intell. 29(11), 1973–1989 (2007)
32. De Jong, K.: Evolutionary Computation: A Unified Approach. MIT Press, Cambridge (2001)
33. Koza, J.R., Keane, M.A., Yu, J., Bennett III, F.H., Mydlowec, W.: Automatic creation of human-competitive programs and controllers by means of genetic programming. Genetic Programming and Evolvable Machines 1(1-2), 121–164 (2000)
34. Poli, R., Langdon, W.B., McPhee, N.F.: A field guide to genetic programming (2008) Published via, `http://lulu.com` and freely available, `http://www.gp-field-guide.org.uk` (With contributions by Koza, J.R.)
35. Olague, G., Cagnoni, S., Lutton, E.: Preface: introduction to the special issue on evolutionary computer vision and image understanding. Pattern Recognition Letters 27(11), 1161–1163 (2006)

36. Cagnoni, S., Lutton, E., Olague., G. (eds.): Genetic and Evolutionary Computation for Image Processing and Analysis. EURASIP Book Series on Signal Processing and Communications, vol. 8. Hindawi Publishing Corporation (2007)
37. Cagnoni, S., Lutton, E., Olague, G.: Editorial introduction to the special issue on evolutionary computer vision. Evoutionary Computation 16(4), 437–438 (2008)
38. Krawiec, K.: Genetic programming-based construction of features for machine learning and knowledge discovery tasks. Genetic Programming and Evolvable Machines 3(4), 329–343 (2002)
39. Ebner, M.: On the evolution of interest operators using genetic programming. In: Poli, R., et al. (eds.) EuroGP 1998. LNCS, vol. 1391, pp. 6–10. Springer, Heidelberg (1998)
40. Zhang, Y., Rockett, P.I.: Evolving optimal feature extraction using multi-objective genetic programming: a methodology and preliminary study on edge detection. In: GECCO 2005: Proceedings of the 2005 conference on Genetic and evolutionary computation, pp. 795–802. ACM, New York (2005)
41. Jaśkowski, W., Krawiec, K., Wieloch, B.: Multitask visual learning using genetic programming. Evol. Comput. 16(4), 439–459 (2008)
42. Lin, Y., Bhanu, B.: Evolutionary feature synthesis for object recognition. IEEE Transactions on Systems, Man and Cybernetics, Part C, Special Issue on Knowledge Extraction and Incorporation in Evolutionary Computation 35(2), 156–171 (2005)
43. Krawiec, K., Bhanu, B.: Visual learning by evolutionary and coevolutionary feature synthesis. IEEE Trans. on Evolutionary Computation 11, 635–650 (2007)
44. Howard, D., Roberts, S.C., Ryan, C.: Pragmatic genetic programming strategy for the problem of vehicle detection in airborne reconnaissance. Pattern Recogn. Lett. 27(11), 1275–1288 (2006)
45. Zhang, M., Ciesielski, V.B., Andreae, P.: A domain-independent window approach to multiclass object detection using genetic programming. EURASIP Journal on Applied Signal Processing, Special Issue on Genetic and Evolutionary Computation for Signal Processing and Image Analysis (8), 841–859 (2003)
46. Song, A., Ciesielski, V.: Texture segmentation by genetic programming. Evol. Comput. 16(4), 461–481 (2008)
47. Keijzer, M., Babovic, V.: Declarative and preferential bias in gp-based scientific discovery. Genetic Programming and Evolvable Machines 3(1), 41–79 (2002)
48. Holland, J.H.: Adaptation in Natural and Artificial Systems. University of Michigan Press, Ann Arbor (1975)
49. Grefenstette, J.J.: Deception considered harmful. In: Whitley, D.L. (ed.) Foundations of Genetic Algorithms 2, pp. 75–91. Morgan Kaufmann, San Mateo (1993)
50. Vose, M.D.: The simple genetic algorithm: foundations and theory. MIT Press, Cambridge (1999)
51. Lindeberg, T.: Feature detection with automatic scale selection. International Journal of Computer Vision 30(2), 79–116 (1998)

Part II
Modular Neural Networks in Pattern Recognition

Face, Fingerprint and Voice Recognition with Modular Neural Networks and Fuzzy Integration

Ricardo Muñoz, Oscar Castillo, and Patricia Melin

Tijuana Institute of Technology, Tijuana México
ocastillo@tectijuana.edu.mx, epmelin@hafsamx.org

Abstract. In this paper we describe a Modular Neural Network (MNN) with fuzzy integration for face, fingerprint and voice recognition. The proposed MNN architecture defined in this paper consists of three modules; face, fingerprint and voice. Each of the mentioned modules is divided again into three sub modules. The same information is used as input to train the sub modules. Once we have trained and tested the MNN modules, we proceed to integrate these modules with a fuzzy integrator. In this paper we demonstrate that using MNNs for face, fingerprint and voice recognition integrated with a fuzzy integrator is a good option to solve pattern recognition problems.

1 Introduction

The identification of people has become a very important activity in different areas of application and with different purposes; in business applications based on punctuality and attendance biometric systems are implemented using the fingerprint or the full hand of the employees; in judicial systems the biometric pattern recognition has become a routine tool in the police force during the criminal investigation, allowing the arrest of criminals worldwide, but also other specific applications are known, such as controlling access to any type of transaction or access to data [1].

Neural networks provide a methodology to work in the field of pattern recognition. Modular neural networks are often used to simplify the problem and obtain good results. Because of this, there arises the need to develop methods to integrate the responses provided by the modules of a modular neural network and thus provide a good result. This paper develops a fuzzy integrator to integrate the responses provided by the modular neural network and achieve a good recognition of persons.

The work described in this paper is divided into two main areas, the first one is about the Modular Neural Network (MNN) and the second one is about the Fuzzy Integrators. The main goal of the first part is to create and test different MNNs architectures for face, fingerprint and voice recognition. The main goal of the

P. Melin et al. (Eds.): Bio-inspired Hybrid Intelligence Systems, SCI 256, pp. 69–79.
springerlink.com

second part is to develop fuzzy integrators to integrate the outputs given by the modules of the modular neural network.

This paper is organized as follows: in sections 2 and 3 we show general neural networks and fuzzy logic theory respectively, in section 4 we describe the problem and mention how to solve it making use of a MNN with fuzzy integration. In sections 4 and 5 we show the results for the MNN trainings and the fuzzy integration respectively. Finally, section 6 shows the conclusions.

2 Neural Networks

Artificial neural networks (NN) are an abstract simulation of a real nervous system that consists of a set of neural units connected to each other via axon connections. These connections are very similar to the dendrites and axons in biological nervous systems. Figure 1 shows the abstract simulation of a real nervous system [2].

Models of artificial neural networks can be classified as:

- Biological models: Networks that try to simulate the biological neural systems, as well as the functions of hearing or some basic functions of vision.
- Models for applications: Models for applications are less dependent on models of biological systems. These are models in which their architectures are strongly linked to the application requirements.

One of the missions of a neural network is to simulate the properties observed in biological neural systems through mathematical models recreated through artificial mechanisms (such as an integrated circuit, a computer or a set of valves). The aim is that the machines give similar responses to those the brain is able to give that are characterized by their robustness and generalization [3].

Artificial neural networks are models that attempt to reproduce the behavior of the brain. As such model, a simplification is made, identifying the relevant elements of the system, either because the amount of information available is excessive or because it is redundant. An appropriate choice of features and an appropriate structure is the conventional procedure used to construct networks capable of performing certain tasks [1].

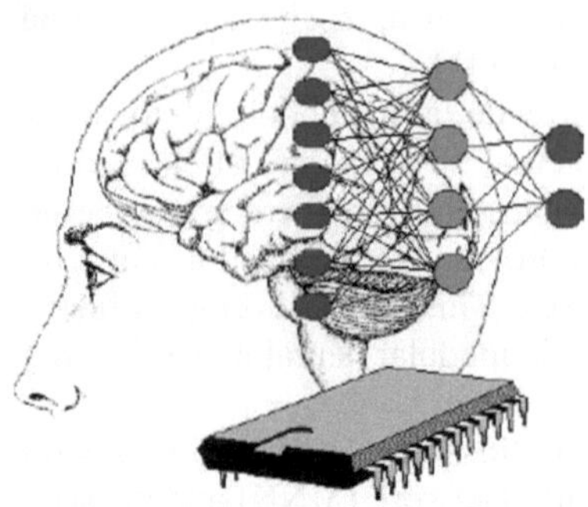

Fig. 1. Simulation of an abstract real nervous system.

Modularity is defined as the ability of a system being studied, seen or understood as the union of several parts interacting with each other and working towards a common goal, each one performing a task necessary to achieve that objective. Each of these parts in which the system is divided is called a module. Ideally, a module must be able to work as a black box, i.e. be independent of other modules and communicate with them (all or only a part) through well-defined inputs and outputs [4].

It is said that a neural network is modular if the computations performed by the network can be decomposed into two or more modules (subsystems) that operate on different inputs with no communication between them. The outputs of the modules are measured by an integration unit, which is not allowed to feed information to the modules. In particular, the unit decides: (1) how the modules are combined to form the final output of the system, and (2) modules that must learn that patterns of training [5].

As mentioned previously, modular neural networks require an integration unit, which allows some form of joining the responses provided by each module. Listed below are some methods that can be used to perform this task

- Average operators
- Gating Network
- *Fuzzy Integrators*
- Voting mechanism using the Softmax function
- Etc.

3 Fuzzy Logic

Fuzzy logic has gained a great reputation as a good methodology for a variety of applications, ranging from control of complex industrial processes to the design of artificial devices for automatic deduction through the construction of electronic

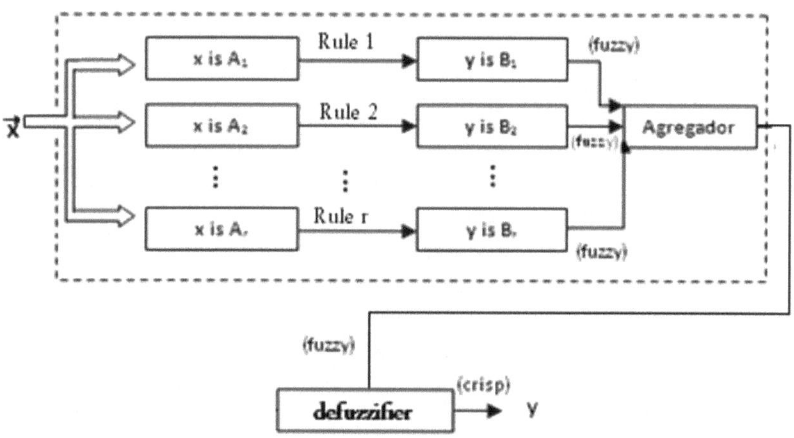

Fig. 2. Fuzzy Inference System.

devices for home use and entertainment, as well as diagnostic systems. It has been considered generally that the concept of fuzzy logic appeared in 1965 at the University of California at Berkeley, introduced by Lotfi A. Zadeh.

The basic structure of a fuzzy inference system consists of three conceptual components: a rule base, which contains a selection of fuzzy rules, a database (or dictionary) which defines the membership functions used in the rules, and a reasoning mechanism that performs the inference procedure [6].

The basic fuzzy inference system can take either fuzzy or traditional inputs, but the outputs it produces are always fuzzy sets. Sometimes you need a traditional output, especially when a fuzzy inference system is used as a controller. So we need a method of "Defuzzification" to extract the numerical value of output (as shown in Figure 2).

4 Modular Neural Network with Fuzzy Integration for Face, Fingerprint and Voice Recognition

The proposed MNN architecture defined in this paper consists of three modules; face, fingerprint and voice. Each of the mentioned modules will be divided into

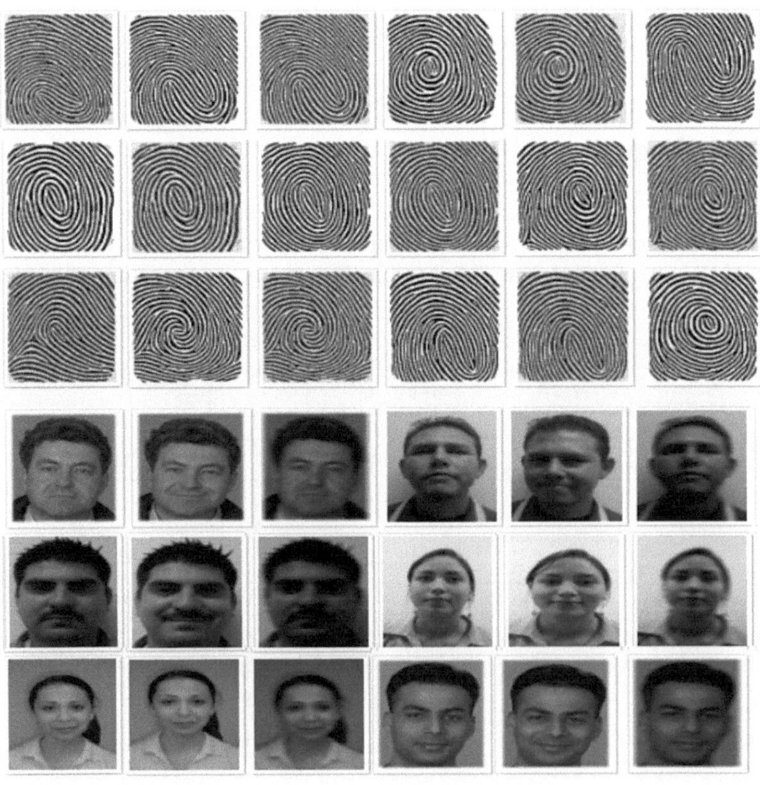

Fig. 3. Fingerprint and face samples.

three sub modules. The same information is used as input to train the sub modules. After the MNN trainings are done, we proceed to integrate the three modules (face, fingerprint ad voice) with a fuzzy integrator.

The data used for training the MNN correspond to 30 persons taken from the ORL database and from students of a master degree in computer science from Tijuana Institute of Technology, Mexico. The details are as follows:

- Face: 90 images taken from [15]. The sizes of these images are of 268x338 pixels with bmp extension and were preprocessed with the Wavelet transform. Two images are used for training (with different gestures) and one with Gaussian noise to test the recognition.
- Fingerprints taken from [15]: 60 fingerprint images. The sizes of these images are of 268x338 pixels with bmp extension and were preprocessed with the Wavelet function. One fingerprint per person for training and one with Gaussian noise to test the recognition.
- Voices taken from [16]: There are 3 word files per person; the words are "hola", "accesar" and "presentacion" in Spanish. The network was trained with the 90 words preprocessed with Mel-cepstral Coefficients and the network is tested with any of them but with noise.

Examples of face and fingerprint are shown in Figure 3 and a general Scheme of the architecture is shown in Figure 4

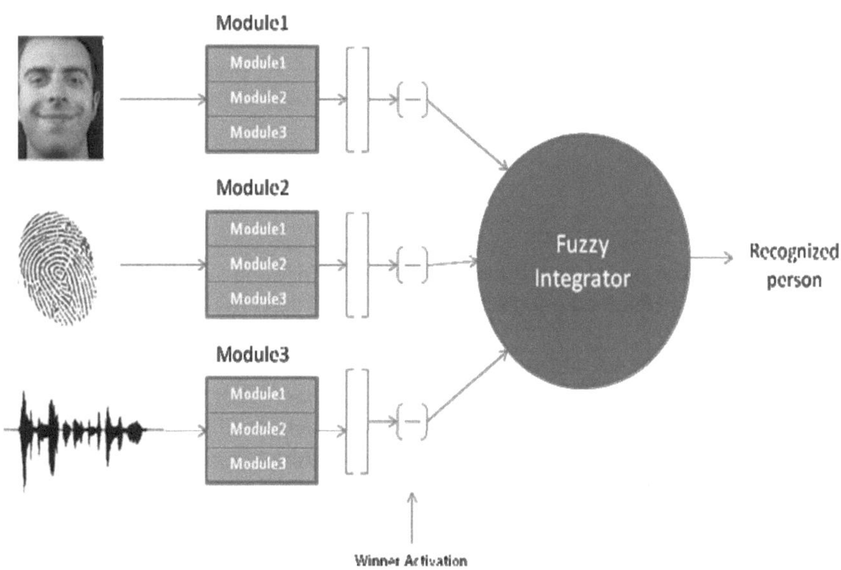

Fig. 4. Architecture of the MNN.

5 Modular Neural Network Results

Several trainings were made separately for each of the modules until we accomplish good results. Tables 1, 2, 3 and 4 show the results of the MNN trainings for face, fingerprint and voice, respectively. Figure 5 shows the results from table 4.

We can observe in the above table that the best result obtained for face module recognition is training ER7 with 100 percent of recognition.

Table 1. Face training results.

Training	Method	Mod	Architecture	Error	Duration	Rec.	%
		1	350,300				
ER1	trainscg	2	350,300	0.01	16:10	19/30	63.33
		3	350,300				
		1	300,300				
ER2	trainscg	2	300,300	0.01	20:34	21/30	70
		3	300,300				
		1	300,240				
ER3	trainscg	2	300,250	0.001	49:39	28/30	93.33
		3	300,260				
		1	350,300				
ER4	trainscg	2	350,300	0.001	40:58	28/30	93.33
		3	350,300				
		1	300,150				
ER5	trainscg	2	300,150	0.001	01:12:30	28/30	93.33
		3	350,150				
		1	290,250				
ER6	trainscg	2	305,210	0.001	50:50	29/30	96.66
		3	320,200				
		1	300,300				
ER7	trainscg	2	300,300	0.001	46:07	30/30	100
		3	300,300				

We can observe in the above table that in several trainings (EH1, EH2, EH3, EH4 and EH5) we obtained good results, having 100 percent of recognition for the fingerprint module.

We can observe in the above tables that the best result obtained for the voice module recognition is training EV1, because it has better performance when different noise levels are applied to the input when testing the module. It is important to know that the noise level consists of adding to each voice data array normally distributed random numbers with a standard deviation of 0.1 to 1 (where noise level 0.1 represents a standard deviation of 0.1, 0.2 a standard deviation of 0.2…1.0 a standard deviation of 1). When the recognized persons equals to 90 we have a 100 percent of recognition.

Table 2. Fingerprint training results.

Training	Method	Mod	Architecture	Error	Duration	Rec.	%
		1	340,175				
EH1	trainscg	2	280,105	0.01	12:51	30/30	100
		3	295,138				
		1	200,100				
EH2	trainscg	2	200,100	0.01	15:56	30/30	100
		3	200,100				
		1	150,80				
EH3	trainscg	2	150,80	0.01	59:34	30/30	100
		3	150,80				
		1	150,100				
EH4	trainscg	2	150,90	0.01	18:09	30/30	100
		3	150,110				
		1	100,50				
EH5	trainscg	2	100,60	0.01	01:39:13	30/30	100
		3	100,55				
		1	100,50				
EH6	trainscg	2	100,60	0.02	16:37	26/30	86.66
		3	100,55				
		1	150,70				
EH7	trainscg	2	117,90	0.02	08:07	26/30	86.66
		3	157,87				

Table 3. Voice training configuration.

Training	Method	Architecture	Error	Duration
		450,240		
EV1	trainscg	420,190	0.001	01:44
		410,225		
		150,80		
EV2	trainscg	170,90	0.001	12:27
		170,75		
		150,100		
EV3	trainscg	150,100	0.001	02:03
		150,100		
		350,140		
EV4	trainscg	320,90	0.001	02:17
		310,125		
		180,80		
EV5	trainscg	180,80	0.001	29:01
		180,80		
		180,120		
EV6	trainscg	180,120	0.001	01:41
		180,120		
		300,100		
EV7	trainscg	300,100	0.001	02:24
		300,100		

Table 4. Voice training results.

	Recognized persons with noise									
Training	0.1	0.2	0.3	0.4	0.5	0.6	0.7	0.8	0.9	1.0
EV1	90	90	90	90	90	90	90	86	81	74
EV2	90	90	90	90	89	86	85	74	62	62
EV3	90	90	90	90	90	88	88	75	50	49
EV4	90	90	90	90	90	90	90	83	78	68
EV5	90	90	90	90	90	89	89	79	64	65
EV6	90	90	90	90	90	89	89	72	64	56
EV7	90	90	90	90	90	89	89	88	81	68

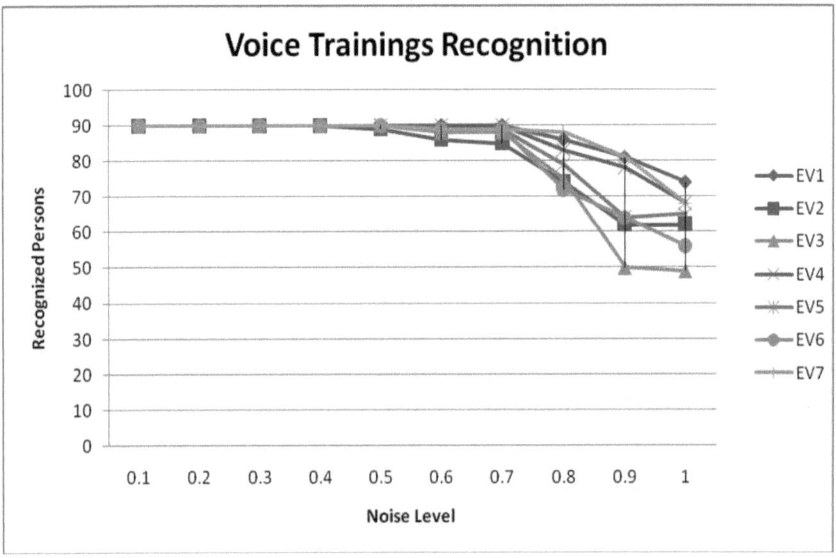

Fig. 5. Results of voice trainings with different noise level.

6 Fuzzy Integration Results

A fuzzy integrator was created using the MATLAB fuzzy logic toolbox, so that we can use it to integrate the MNN modules and generate a final result. This fuzzy integrator consists of 27 rules; three inputs that are provided by the MNN: face activation, fingerprint activation and voice activation; one output: winner

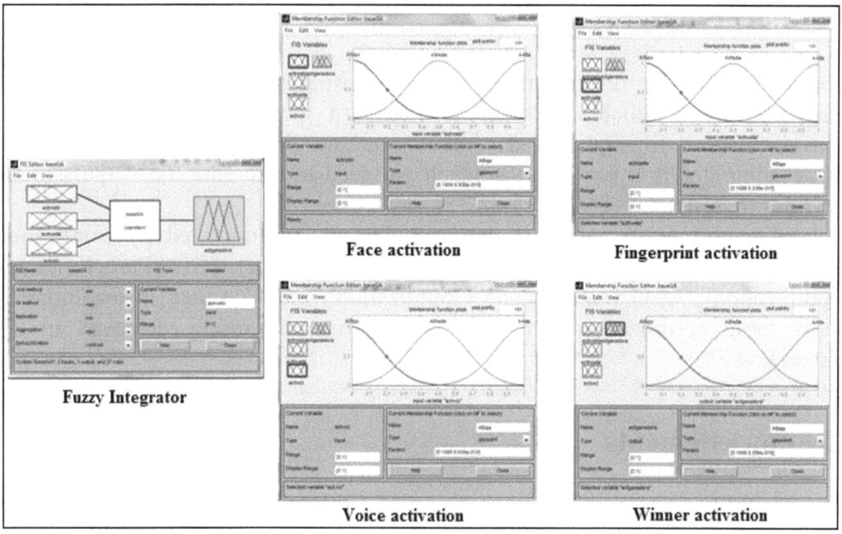

Fig. 6. Architecture of the fuzzy integrator.

Table 5. Fuzzy integration results.

No.	Face	Rec.	Finger print	Rec.	Voice-noise		Rec.	Integration
1	ER1	63.33%	EH6	86.66%	EV3	1.0	49%	71/90(78.89%)
2	ER1	63.33%	EH1	100%	EV3	1.0	49%	90/90 (100%)
3	ER7	100%	EH6	86.66%	EV3	1.0	49%	80/90(88.89%)
4	ER7	100%	EH1	100%	EV3	1.0	49%	90/90 (100%)
5	ER1	63.33%	EH6	86.66%	EV1	0.5	100%	73/90(81.10%)
6	ER1	63.33%	EH1	100%	EV1	0.5	100%	90/90(100%)
7	ER7	100%	EH6	86.66%	EV1	0.5	100%	81/90(90%)
8	ER7	100%	EH1	100%	EV1	0.5	100%	90/90(100%)

activation that will indicate which person is recognized. It is important to notice that each input and output has three Gaussian membership functions. Figure 6 shows the architecture of the fuzzy integrator.

Table 5 below shows the results from the fuzzy integration of the MNN. These results were obtained by making some combinations of the MNN modules. We can observe that when all three modules have 100% of recognition, fuzzy integration is perfect (i.e. No. 8). Even in some cases, where one or two of the modules does not give good recognition, the fuzzy integration is perfect too (i.e. no.4, 6 and 7). But, when all three modules have poor recognition, the fuzzy integration is not good (i.e. no. 1).

7 Conclusions

The analysis of the results presented in this paper shows that Modular Neural Networks are a reliable technique for pattern recognition. In this case we considered face, fingerprint and voice as patterns. We also demonstrated that a fuzzy integrator is a good option when we need to integrate the outputs of the modules of a MNN.

It is important to note that these results can be improved by using some optimization techniques, like a GA (Genetic Algorithm), PSO (Particle Swarm Optimization), ACO (Ant Colony Optimization), etc. We plan in the future to apply these optimization techniques to improve the results achieved in this research work.

Acknowledgments

We would like to express our gratitude to the CONACYT and Tijuana Institute of Technology for the facilities and resources granted for the development of this research.

References

[1] Artificial neural networks fundamentals, models and applications (December 2008), http://www.monografias.com/trabajos12/redneur/redneur.shtml
[2] Connectionist systems (December 2008), http://carpanta.dc.fi.udc.es/~cipenedo/cursos/scx/archivospdf/Tema1-0.pdf
[3] Artificial neural networks (December 2008), http://es.wikipedia.org/wiki/Red_neuronal_artificial
[4] Modularity (January 2009), http://es.wikipedia.org/wiki/Modularidad
[5] Kuri, F.: Neural Networks and Genetic Algorithms. PDF document
[6] Jang, J.S.R., Sun, C.T., Mizutani, E.: Neuro-Fuzzy and Soft Computing. Prentice Hall, New Jersey (1997)
[7] Ripley, B.D.: Pattern Recognition and Neural Networks. Cambridge University Press, Cambridge (1996)
[8] Morales, G.: Introduction to Fuzzy Logic, Research center and Advanced studies of the IPN, http://delta.cs.cinvestav.mx/~gmorales/ldifll/node1.html
[9] Castro, J.R.: Tutorial Type-2 Fuzzy Logic: Theory and Applications. UABC University and Tijuana Institute of Technology, Tijuana México (2006)
[10] Fuzzy Logic: Introduction and basic concepts (January 2009), http://members.tripod.com/jesus_alfonso_lopez/FuzzyIntro2.html
[11] Fundamentals of Fuzzy Logic (January 2009), http://www.itq.edu.mx/vidatec/espacio/aiee/fuzzy.ppt

[12] Genetic Algorithms (December 2008), http://es.wikipedia.org/wiki/Algoritmo_gen%C3%A9tico
[13] Golberg, D.: Genetic Algorithms in search, optimization and machine learning. Addison Wesley, Reading (1989)
[14] Langari, R.: A Framework for analysis and synthesis of fuzzy linguistic control systems. Ph.D. thesis. University of California, Berkeley (1990)
[15] Alvarado, J.M.: Recognition of the person through his face and fingerprint using modular neural networks and wavelet transform, Tijuana Institute of Technology, Tijuana México (2006)
[16] Ramos, J.: Neural networks applied to speaker identification by voice using feature extraction, Tijuana Institute of Technology, Tijuana México (2006)

Modular Neural Networks with Fuzzy Response Integration for Signature Recognition

Mónica Beltrán, Patricia Melin, and Leonardo Trujillo

Tijuana Institute of Technology, Tijuana, México
epmelin@hafsamx.org

Abstract. This chapter describes a modular neural network (MNN) for the problem of signature recognition. Currently, biometric identification has gained a great deal of research interest within the pattern recognition community. For instance, many attempts have been made in order to automate the process of identifying a person's handwritten signature, however this problem has proven to be a very difficult task. In this work, we propose a MNN that has three separate modules, each using different image features as input, these are: edges, wavelet coefficients, and the Hough transform matrix. Then, the outputs from each of these modules are combined using a Sugeno fuzzy integral. The experimental results obtained using a database of 30 individual's shows that the modular architecture can achieve a very high 98% recognition accuracy with a test set of 150 images. Therefore, we conclude that the proposed architecture provides a suitable platform to build a signature recognition system.

1 Introduction

Recently, there has been an increased interest in developing biometric recognition system for security and identity verification purposes [2]. Such systems usually are intended to recognize different types of human traits, which include a person's face, their voice, fingerprints, and specific handwriting traits [2].

Particularly, the handwritten signature that each person posses is widely used for personal identification and has a rich social tradition. In fact, currently it is almost always necessary in all types of transactions that involve legal or financial documents.

However, it is not a trivial task for a computational system to automatically recognize a person's signature for the following reasons. First, there can be a great deal of variability when a person signs a document. This can be caused by different factors, such as a person's mood, free time to write the signature, and the level of concentration during the actual act of signing a document. Second, because signatures can be so diverse it is not evident which type of features should be used in order to describe and effectively differentiate among them. For instance, some signatures are mostly written using straight line segments, and still others have

P. Melin et al. (Eds.): Bio-inspired Hybrid Intelligence Systems, SCI 256, pp. 81–91.
springerlink.com

a much smoother form with curved and circular lines. Finally, many signatures share common traits that make them appear quite similar depending on the types of features that are analyzed.

In this work, we present a handwritten signature recognition system using Modular Neural Networks (MNNs) with a Sugeno fuzzy integral. We have chosen a MNN because they have proven to be a powerful, robust, and flexible tool, useful in many pattern recognition problems [12, 13]. In fact, we only extract simple and easily computed image features during our preprocessing stage, these features are: image edges, wavelet transform coefficients, and the Hough transform matrix. The MNN we propose uses these features to perform a very accurate discrimination of the input data used in our experimental tests. Therefore, we have confirmed that a MNN system can solve a difficult biometric recognition problem using a simple set of image features.

2 Problem Statement and Outline of Our Proposal

The problem we address in this chapter is concerned with the automatic recognition of a person's signature that is captured on a Tablet PC. We suppose that we have a set of N different people, and each has a unique personal signature. The system is trained using several samples from each person, and during testing it must determine the correct label for a previously unknown sample.

The system we are proposing consists on a MNN with three separate modules. Each module is given as input the features extracted with different feature extraction methods: edge detection, wavelet transform, and Hough transform. The responses from each of the modules are combined using a Sugeno fuzzy integral which determines the person to whom the input signature corresponds. A general schematic of this architecture is shown in Figure 1, where all of the modules and stages are clearly shown.

In the following section we present a brief review of some of the main concepts needed to understand our work.

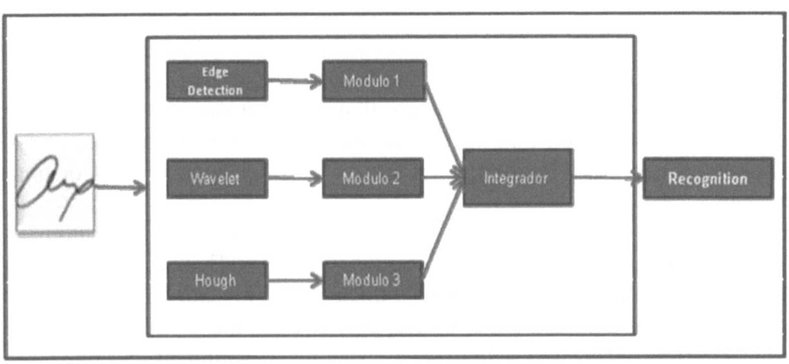

Fig. 1. General architecture of the proposed Modular Neural Network for signature recognition.

3 Background Theory

In this section we provide a general review of artificial neural networks and modular architectures, we discuss how output from the modular system can be integrated using Sugeno fuzzy integrals, and we describe the feature extraction methods that provide the input for each of the modules in our MNN.

3.1 Modular Neural Networks

Artificial Neural Networks (ANNs) are information processing systems that employ a conceptual model that is based on the basic functional properties of biological neural networks. In the past twenty or thirty years, ANN research has grown very rapidly, in the development of new theories of how these systems work, in the design of more complex and intricate models, and in their application to a diverse set problem domains. Regarding the latter, application domains for ANN include pattern recognition, data mining, time series prediction, robot control, and in the development of hybrid methods with fuzzy logic and genetic algorithms, to mention but a few examples [6,12, 13].

In canonical implementations, most systems employ a monolithic network in order to solve the given task. However, when a system needs to process large amounts of data or when the problem is highly complex, then it is not trivial, and sometimes unfeasible, to establish a good architecture and topology for a single network that can solve the problem. For instance, in such problems a researcher might attempt to use a very large and complex ANN. Nevertheless, large networks are often difficult to train, and for this reason they rarely achieve the desired performance [12].

In order to overcome some of the aforementioned shortcomings of monolithic ANNs, many researchers have proposed modular approaches [11]. MNNs are based on the general principle of divide-and-conquer, where one attempts to divide a large problem into smaller sub-problems that are easier to solve independently. Then, these partial solutions are combined in order to obtain the complete solution for the original problem.

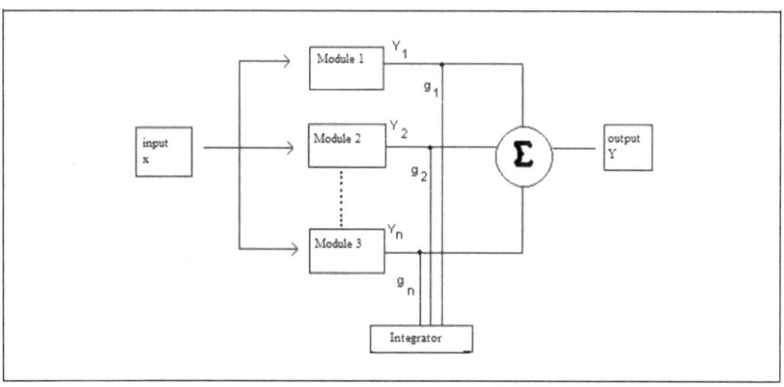

Fig. 2. Architecture of a Modular Network.

MNNs employ a parallel combination of several ANNs, and normally contain two main components: (1) local experts; and (2) an integrating unit. The basic architecture is shown in Figure 1 [14].

Each module consists of a single ANN, and each is considered to be an expert in a specific task. After the input is given to each module it is necessary to combine all of the outputs in some way, this task is carried out by a special module called an *integrator*. The simplest form of integration is given by a gating network, which basically switches between the outputs of the different modules based on simple criteria, such as the maximum level of activation. However, a better combination of the response from each module can be obtained using more elaborate methods of integration, such as the Sugeno fuzzy integral [14].

3.2 Sugeno Fuzzy Integral

The Sugeno fuzzy integral is a nonlinear aggregation operator that can combine different sources of information [3, 7, 8]. The intuitive idea behind this operator is based on how humans integrate information during a decision making process. In such scenarios it is necessary to evaluate different attributes, and to assign priorities based on partially subjective criteria. In order to replicate this process on an automatic system, a good model can be obtained by using a fuzzy representation [4, 7, 11]. Finally, several works have shown that the use of a Sugeno fuzzy integral as a MNN integrator can produce a very high level of performance [5, 7, 10], and for these reasons we have chosen it for the system we describe here.

3.3 Feature Extraction

In this work, we employ three individual modules, and each receives different image features extracted from the original image of a person's signature. Each of these feature extraction methods are briefly described next.

3.3.1 Edge Detection

For images of handwritten signatures edges can capture much of the overall structure present within, because people normally write using a single color on a white

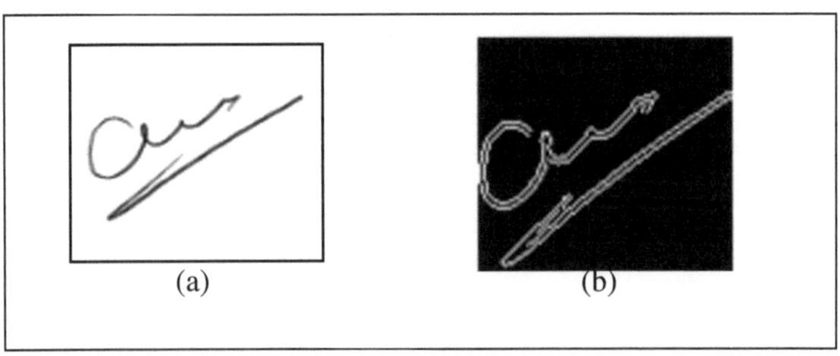

Fig. 3. (a) Original image of a signature. (b) Image edges.

background. Hence, we have chosen to apply the Canny edge detector to each image that generates a binary image of edge pixels, see Fig 3.

3.3.2 Wavelet Transform

The wavelet transform decomposes a signal using a family of orthogonal functions, it accounts for both the frequency and the spatial location at each point. The most common application is the Discrete Wavelet Transform (DWT) using a Haar wavelet [15]. The DWT produces a matrix of wavelet coefficients that allows us to compress, and if needed reconstruct, the original image. In Figure 4 we can observe the two compression levels used in our work.

3.3.3 Hough Transform

In the third and final module we employ the Hough transform matrix as our image features [1]. The Hough transform can extract line segments from the image. In Figure 5 we show a sample image of a signature and its corresponding Hough transform matrix. Finally, in order to reduce the size of the matrix, and the size of the corresponding ANN, we compress the information of the Hough matrix by 25%.

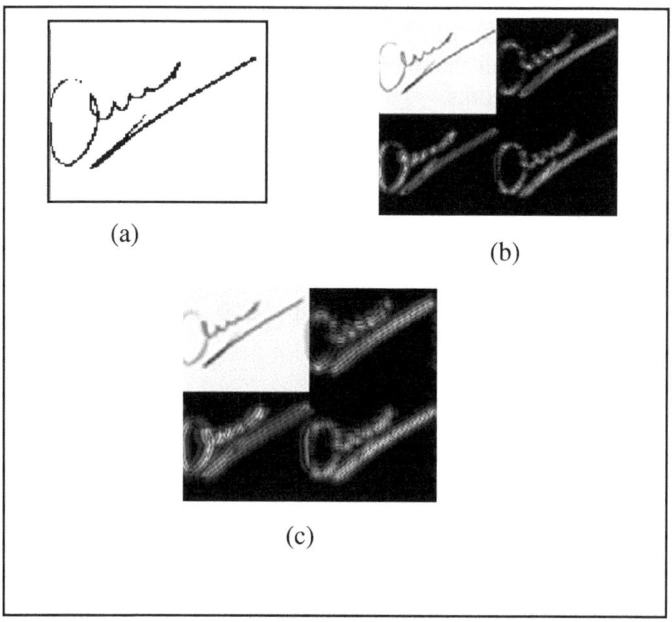

Fig. 4. (a) Original image of a signature. (b) First level of decomposition. (c) Second level of decomposition.

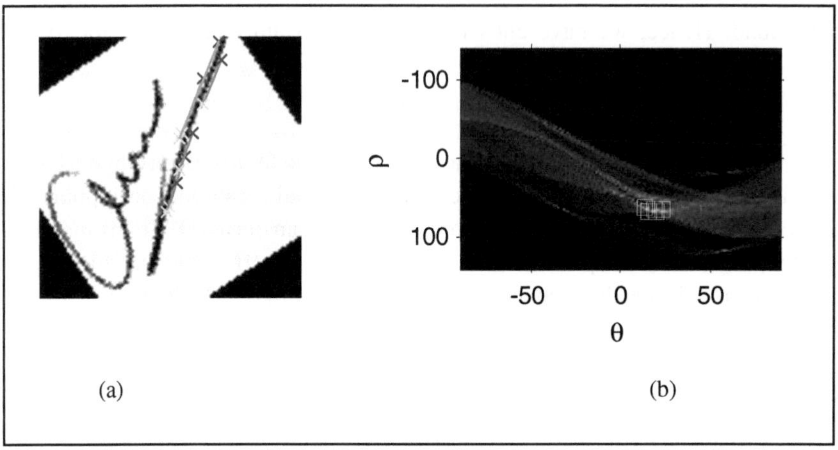

Fig. 5. (a) Sample of a signature image with some of the lines found by the Hough transform (b) The Hough transform matrix

4 Experiments

In this section we present our database of signature images, describe our experimental set-up, and detail the experimental results we have obtained using monolithic and modular networks.

4.1 Image Database

For this work we build a database of images with the signatures of 30 different people, students and professors from the computer science department at the Tijuana Institute of Technology, BC, México. We collected 12 samples of the signature from each person; this gives a total of 360 images in total. Sample images from the database are shown in Figure 6.

4.2 Experimental Setup

In this work, we are interested in verifying the performance of our proposed MNN for the problem of signature recognition. Therefore, in order to obtain comparative measures we divide our experiments into two separate tests.

1. First, we use each module as a monolithic ANN for signature recognition. Therefore, we obtain three sets of results, one for each module, where in each case a different feature extraction method is used.
2. Second, we train our MNN using all three modules concurrently and the Sugeno fuzzy integral as our integration method.

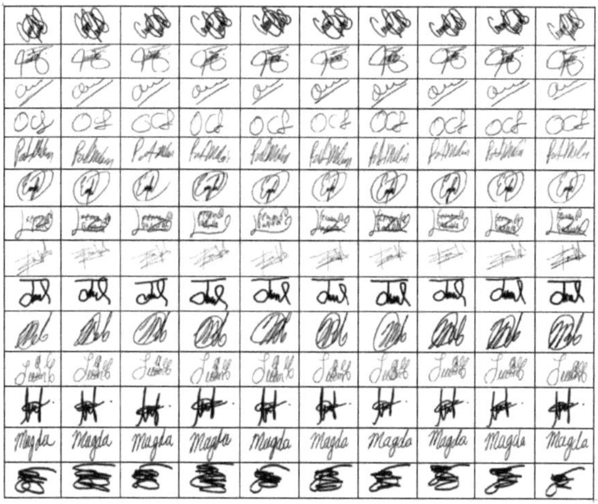

Fig. 6. Images from our database of signatures. Each row shows different samples from the signature of the same person.

In all tests 210 images were chosen randomly and used for training, and the remaining 150 were used as a testing set. Additionally, after some preliminary runs it was determined that the best performance was achieved when the ANNs were trained with the Scaled Conjugate Gradient (Trainscg) algorithm, with a goal error of 0.001. Moreover, all networks had the same basic ANN architecture, with two hidden layers. In what follows, we present a detailed account of each of these experimental tests.

4.2.1 Monolithic ANNs

The results for the first monolithic ANN are summarized in Table 1. The table shows a corresponding ID number for each training case, the total epochs required to achieve the goal error, the neurons in each hidden layer, and the total time required for training. Recognition performance is shown with the number of correct recognitions obtained with the 150 testing images, and the corresponding accuracy score. In this case, the best performance was achieved in the third training run where the algorithm required 78 epochs, and the ANN correctly classified 131 of the testing images.

The second monolithic ANN uses the wavelet features as input, and the obtained results are summarized in Table 2. In this case the best performance was obtained in the third training run, with a total of 5 epochs, and 144 correctly classified images. It is obvious that wavelet features provide a very good discriminative description of the signature images that we are testing.

Table 1. Performance for a monolithic ANN using edge features; bold indicates best performance.

No	Epochs	Neurons	Time	Correct	Accuracy (%)
01	80	100-100	00:01:11	123/150	82
02	59	100-100	00:01:18	120/150	80
03	**78**	**100-100**	**00:01:07**	**131/150**	**87**
04	90	100-100	00:01:26	117/150	78
05	80	100-100	00:01:08	119/150	79
06	78	100-100	00:01:34	123/150	82
07	53	100-100	00:00:46	123/150	82
08	79	80-90	00:01:08	123/150	82
09	55	80-90	00:00:56	128/150	85
10	58	80-90	00:00:58	122/150	81

Table 2. Performance for a monolithic ANN using wavelet features.

Train	Epochs	Neurons	Time	Correct	Accuracy (%)
01	12	100-100	00:00:18	135/150	90
02	30	100-100	00:00:25	138/150	92
03	**05**	**100-100**	**00:00:08**	**144/150**	**96**
04	09	100-100	00:00:11	140/150	93
05	06	80-90	00:00:08	142/150	95
06	05	80-90	00:00:05	140/150	93
07	10	80-90	00:00:14	141/150	94
08	07	80-90	00:00:09	138/150	92
09	10	80-90	00:00:15	140/150	93
10	05	80-90	00:00:06	137/150	91

Finally, the third monolithic ANN uses the Hough transform matrix, and the corresponding results are shown in Table 3. The best performance is achieved in the fourth training run, with a total of 6 epochs and 141 correctly classified images.

It is important to note that in all three cases, the monolithic methods did achieve good results. The best performance was obtained using wavelet features, and the Hough transform matrix also produced very similar results. On the other hand, the simple edge features produced a less accurate recognition than the other two methods.

Table 3. Performance for a monolithic ANN using the Hough transform.

Train	Epochs	Neurons	Time	Correct	Accuracy (%)
01	63	100-100	00:00:19	135/150	90
02	65	100-100	00:00:51	140/150	93
03	68	100-100	00:00:19	141/150	94
04	**06**	**80-90**	**00:00:08**	**141/150**	**94**
05	04	80-90	00:00:05	140/50	93
06	45	80-90	00:00:11	138/150	92
07	08	80-90	00:00:09	138/150	92
08	05	80-90	00:00:08	137/150	91
09	05	80-90	00:00:06	137/150	91
10	33	50-50	00:00:20	138/150	92

4.2.2 Modular Neural Network with Sugeno Fuzzy Integral

The final experimental results correspond to the complete MNN described in Figure 1, and Table 4 summarizes the results of ten independent training runs. For the modular architecture, performance was consistently very high across all runs, and the best recognition accuracy of 98% was achieved in half of the runs. In fact, even the worst performance of 95% is better or equal than all but one of the monolithic ANNs (see Table 2).

Table 4. Results for the Modular Neural Network

Trian	Epochs	Time	Correct	Accuracy (%)
01	**55**	**00:00:49**	**147/150**	**98**
02	150	00:01:34	146/150	97
03	180	00:01:53	144/150	96
04	300	00:02:20	147/150	98
05	150	00:01:30	147/150	98
06	155	00:01:45	146/150	97
07	320	00:02:49	147/150	98
08	310	00:02:38	148/150	98
09	285	00:01:58	145/150	96
10	03	00:00:02	143/150	95

5 Summary, Conclusions and Future Work

In this chapter we have addressed the problem of signature recognition, a common behavioral biometric measure. We proposed a modular system using ANNs and three types of image features: edges, wavelet coefficients, and the Hough transform matrix. In our system, the responses from each module were combined using a Sugeno fuzzy integral. In order to test our system, we built a database of image signatures from 30 different individuals. In our experiments, the proposed architecture achieves a very high recognition rate, results that confirm the usefulness of the proposal.

In our tests, we have confirmed that the modular approach always outperforms, with varying degrees, the monolithic ANNs tested here. However, in some cases the difference in performance was not very high, only 3 or 2 percent. Nevertheless, we believe that if the recognition problem is made more difficult than the modular approach will more clearly show a better overall performance.

Furthermore, our results also show that even with the simple image features used in this work, each of the ANN modules is indeed capable of learning very good discriminating functions that can correctly differentiate between our set of image signatures.

Finally, the results we have obtained suggest several possible extensions for our work, which include the following:

1. Test the system with a more challenging image database, using more signatures and a smaller set of training samples, in order to verify the robustness of our approach.
2. Employ different integration methods with the MNN architecture described here; e.g., a Fuzzy Inference System.
3. Employ the MNN approach to solve a closely related task, such as signature verification which implies a more difficult classification problem.

References

[1] Ballard, D.: Generalizing the Hough transform to detect arbitrary shapes. Pattern Recognition 13(2), 111–122 (1981)

[2] Zhang, D.: Automated Biometrics Technologies and Systems, Hong Kong Polytechnic University, ch. 10, pp. 203–206. Kluwer Academic Publishers, Dordrecht (2000)

[3] Keller, J., Gader, P., Hocaoglu, A.: Integrals in image processing and recognition. In: Grabisch, M., et al. (eds.) Fuzzy Measures and Integrals: Theory and Applications, pp. 435–466. Physica-Verlag, NY (2000)

[4] Grabisch, M.: A new algorithm for identifying fuzzy measures and its application to pattern recognition. In: Proc. of 4th IEEE Int. Conf. on Fuzzy Systems, Yokohama, Japan, pp. 145–150 (1995)

[5] Grabisch, M., Murofushi, T., Sugeno, M.: Fuzzy Measures and Integrals: Theory and Applications, pp. 348–373. Physica-Verlag, NY (2000)

[6] Castillo, O., Melin, P., Kacprzyk, J., Pedrycz, W.: Hybrid Intelligent Systems Analysis and Design. Springer, Heidelberg (2007)

[7] Mendoza, O., Melin, P.: The Fuzzy Sugeno Integral as a Decision Operator in the Recognition of Images with Modular Neural Networks, Tijuana Institute of Technology, México. Springer, Heidelberg (2007)
[8] Melin, P., Mancilla, A., González, C., Bravo, D.: Modular Neuronal Networks with Fuzzy Sugeno Integral Response Integration for Face and Fingerprint Recognition. In: The International MultiConference in Computer Science and Computer Enginnering, Las Vegas, USA, vol.1, pp. 91–97 (2004)
[9] Melin, P., Mancilla, A., Lopez, M., Solano, D., Soto, M., Castillo, O.: Pattern Recognition for Industrial Security using the Fuzzy Sugeno Integral and Modular Neural Network. In: Soft Computing in Industrial Applications, vol. 39, pp. 4–9. Springer, Heidelberg (2007)
[10] Melin, P., Felix, C., Castillo, O.: Face Recognition using Modular Neural Networks and the Fuzzy Sugeno Integral for Response Integration. Journal of Intelligent Systems 20(2), 275–29 (2005)
[11] Melin, P., Gonzalez, C., Bravo, D., Gonzalez, F., Martinez, G.: Modular Neural Networks and Fuzzy Sugeno Integral for Pattern Recognition: Then Case of Human Face and Fingerprint, Tijuana Institute of Technology, Tijuana, Mexico. Springer, Heidelberg (2007)
[12] Melin, P., Castillo, O.: Hybrid Intelligent Systems for Pattern Recognition Using Soft Computing: An Evolutionary Approach for Neural Networks and Fuzzy Systems, 1st edn. Studies in Fuzziness and Soft Computing. Springer, Heidelberg (2005)
[13] Melin, P., Castillo, O.: Hybrid Intelligent Systems for Pattern Recognition. Springer, Heidelberg (2005)
[14] Kung, S., Mak, M., Lin, S.: Biometric Authentication A Machine Learning Approach. Prentice Hall Information and System Sciences Series (Kailath, T., Series ed.), pp. 27–49 (2005)
[15] Santoso, S., Powers, E., Grady, E.: Power quality disturbance data compression using wavelet transform methods. IEEE Trans. On Power Delivery 12(3), 1250–1257 (1997)

Intelligent Hybrid System for Person Identification Using Biometric Measures and Modular Neural Networks with Fuzzy Integration of Responses

Magdalena Serrano, Erika Ayala, and Patricia Melin

Tijuana Institute of Technology, Tijuana México
epmelin@hafsamx.org

Abstract. This paper presents an intelligent system for person identification with biometric measures such as signature, fingerprint and face. We describe the neural network architectures used to achieve person identification based on the biometrics measures. Simulation results show that the proposed method provides good recognition.

1 Introduction

At the moment, systems based on biometric recognition have gained importance in systems that require the identification of users or restricted access. Compared with conventional methods used as keys, we have the advantage that the biometric features may not be provided, copied or stolen.

These kinds of systems are usually easy to maintain. A biometric system is essentially a pattern recognition system that operates in the following manner: capturing a biometric measure, a set of features are extracted and compared with another group the features.

Biometric identification techniques are very diverse, since any element of a person is potentially usable as a biometric measure. Even with the diversity of existing techniques, to develop a biometric identification system, it is an entirely separate scheme from the technique used. The most used biometric measures are: fingerprint, iris, voice, signature, face, ear, hand geometry, vein structure, retina, etc.

The need for a way to identify the human being in a unique way, has led men to implement a wide range of methods.

Until today biometric methods have been implemented using different devices for the creation of patterns and generate the code that identifies the individual biometric. This is why, in this work we consider one of the most used biometric methods throughout history, which is the fingerprint, and other ones such as the face and the signature.

P. Melin et al. (Eds.): Bio-inspired Hybrid Intelligence Systems, SCI 256, pp. 93–109.
springerlink.com

2 Neural Networks

A neural network is a model to perform a computational simulation of parts of the human brain using replication behavior in small-scale patterns that it performs for producing results from the perceived events.

An artificial neural network (ANN), often just called a "neural network" (NN), is a mathematical model or computational model based on biological neural networks. It consists of an interconnected group of artificial neurons and processes information using a connectionist approach to computation. In most cases an ANN is an adaptive system that changes its structure based on external or internal information that flows through the network during the learning [2].

1. Biological neural networks are made up of real biological neurons that are connected or functionally related in the peripheral nervous system or the central nervous system. In the field of neuroscience, they are often identified as groups of neurons that perform a specific physiological function in laboratory analysis.
2. Artificial neural networks are made up of interconnecting artificial neurons (programming constructs that mimic the properties of biological neurons). Artificial neural networks may either be used to gain an understanding of biological neural networks, or for solving artificial intelligence problems without necessarily creating a model of a real biological system. The real, biological nervous system is highly complex and includes some features that may seem superfluous based on an understanding of artificial networks.

2.1 Structure of an Artificial Neural System

The artificial neural system is composed of several components that are necessary to structure the system. In figure 1 we show these components: the neuron, layers, and networks.

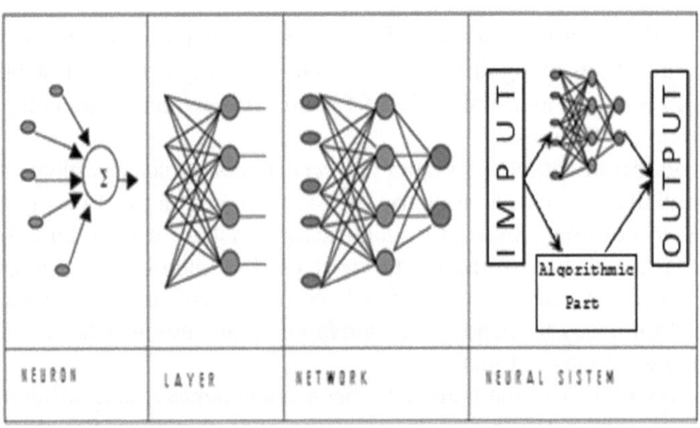

Fig. 1. Hierarchical structure of a system based on artificial neural networks

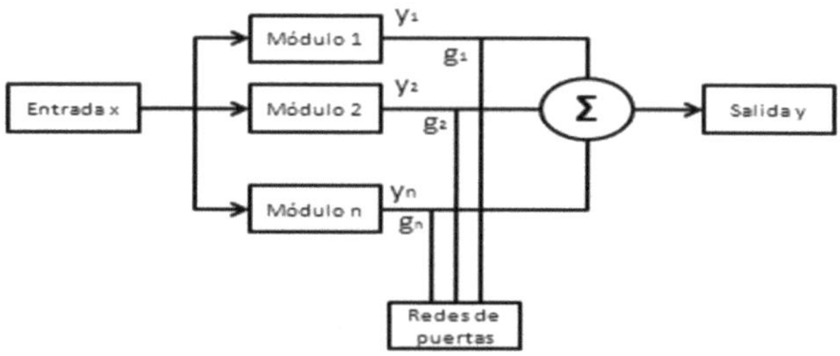

Fig. 2. A modular neural networks architecture.

2.2 Modular Neural Networks

A modular neural network is a neural network characterized by a series of independent neural networks moderated by some intermediary. Each independent neural network serves as a module and operates on separate inputs to accomplish some subtask of the task the network is intended to perform. The intermediary takes the outputs of each module and processes them to produce the output of the network as a whole. The intermediary only accepts the modules' outputs—it does not respond to, nor otherwise signal, the modules. Also the modules do not interact with each other [4].

The advantage is that if the model supports naturally a breakdown into more simple functions, the application of a modular network translates into faster learning. Each module can be built differently, in a way that meets the requirements of each subtask. A modular neural network can be represented by a scheme shown in figure 2.

3 Characteristics of a Biometric Measure

A biometric measure is a feature that can be used to make an identification. Whatever the measure, it must meet the following requirements:

1. Universality: means that anyone should have that characteristic.
2. Uniqueness: the existence of two people with identical characteristics has a very small probability.
3. Permanent Characteristics: the characteristic does not change over time
4. Quantification: the characteristic can be measured in a quantitative form.

3.1 Architecture of a Biometric System for Personal Identification

A biometric system has three basic components: The first is responsible for the acquisition of any analog or digital biometric feature of a person, such as the acquisition of a fingerprint image using a scanner. The second handles the compression, processing, storage and comparison of data acquired with the stored data. The third component provides an interface to applications on the same or another system. In figure 3 we show the phases of a biometrical identification system.

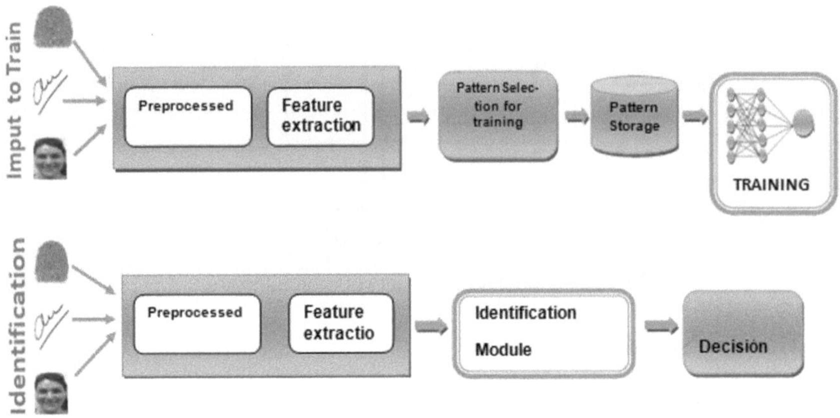

Fig. 3. Phases of a biometrical identification system.

4 Problem Statement and Proposed Method

The problem is to develop a hybrid system for person identification using parallel processing and with biometric measures. Where we must develop a module for each of the biometric measures in the neural network architecture to implement:

- Fingerprint.
- Signature.
- Face.

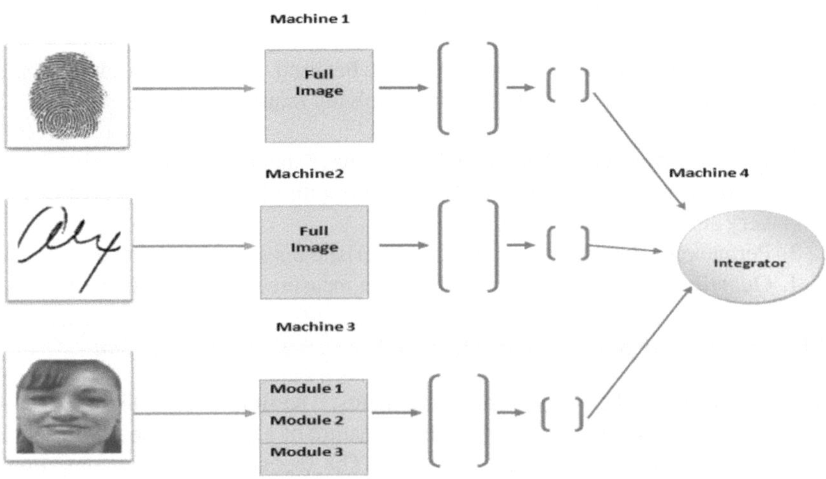

Fig. 4. Proposed scheme for the development of the system.

Starting from the 3 above-mentioned architectures it is important to make the unification of the 3 biometric measures to make the system obtain a higher percentage of identification, which aims at making the integration of the 3 systems into one, as shown in Figure 4.

The implementation was done in Quad core computers, i.e, with 4 computers with 4 processors to active 16 processors in parallel.

4.1 Biometric Databases

Databases were obtained from students and professors of the Master in science in computer science from Tijuana Institute the Technology.

The database consists of 10 Samples of signature, fingerprint and face, taken from 30 people, giving a total of 300 samples. We used for training the first 7 images and leaving the last 3 for identification. In total we trained the network with 210 images and 90 are left for the identification for face, fingerprint and signature.

The databases are shown in figure 6, 9, 10 and 11.

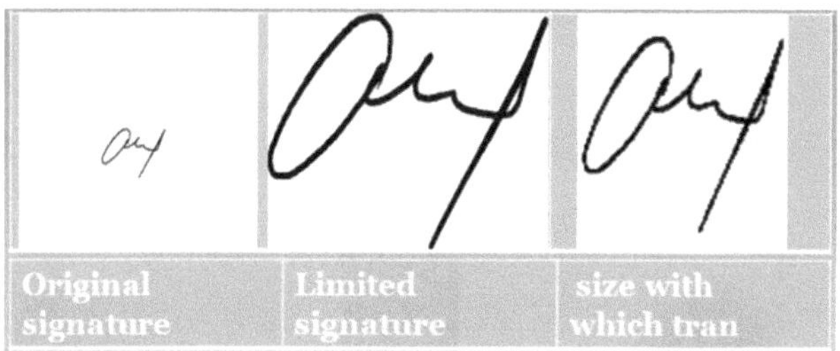

Fig. 5. Preprocessed database

Fig. 6. Database with examples of signatures.

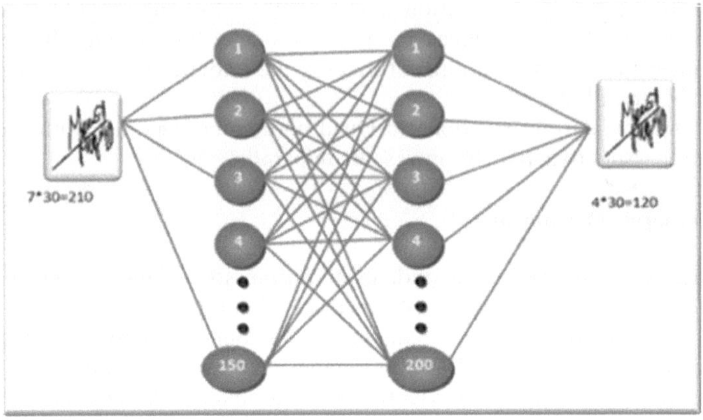

Fig. 7. Architecture of the net for signature.

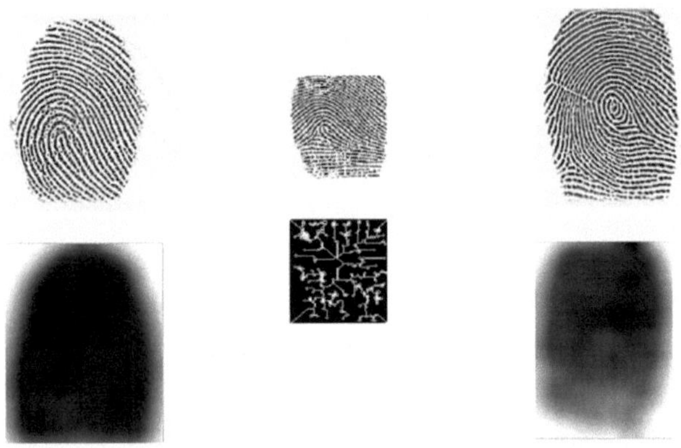

Fig. 8. Preprocessing of fingerprint.

4.1.1 Signature Database

For the normalization of this database the images were cut to limit the signature, this produced resizing of the images. We also applied Wavelets and the Sobel operator to preprocess the image, as shown in Figure 5.

After the preprocessing, the database can be visualized in Figure 6 with examples of signatures .

The architecture for training the net is shown in Figure 7, were we have 2 hidden layers.

4.1.2 Fingerprint Database

For the normalization of the database, a cut of the standard image of the limitation of signature was performed, preprocessing was also performed, and the methods used for this are the following:

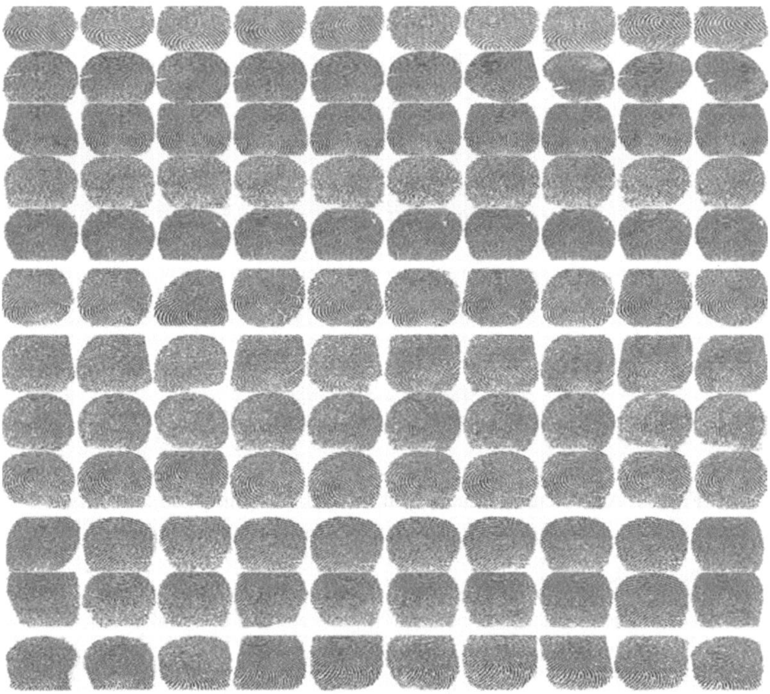

Fig. 9. Database of fingerprints

- Gradient magnitude
- Sobel edge masks
- Skeleton.
- Wavelets.

In figure 8 we show an example of fingerprint preprocessing. We show the gradient magnitude and skeleton methods applied to the fingerprint.

After preprocessing the database of images, they go in to the network with a dimension of [75 * 50] as shown in Figure 9.

The architecture for training the net is shown in figure 10 and it has 2 hidden layers.

4.1.3 Face Database

The face data base consist of a mixture of the ORL and data acquired with students and professors of the institution, and has a different treatment for this reason, we first converted the images to gray tones and made a resizing to [92*112] in BMP format. The face database of the institution is shown en figure 11.

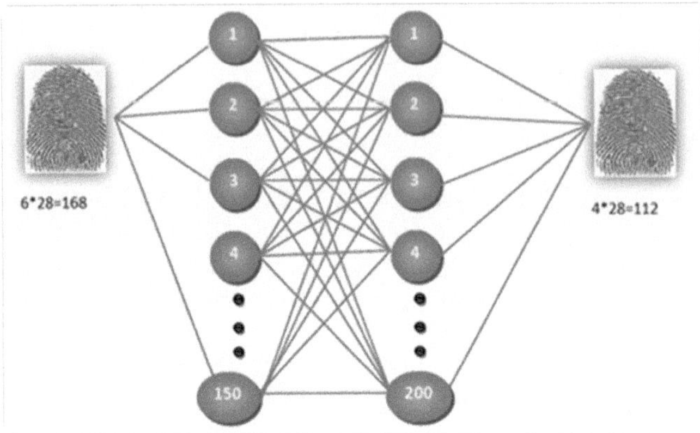

Fig. 10. Architecture of the neural network for fingerprint recognition.

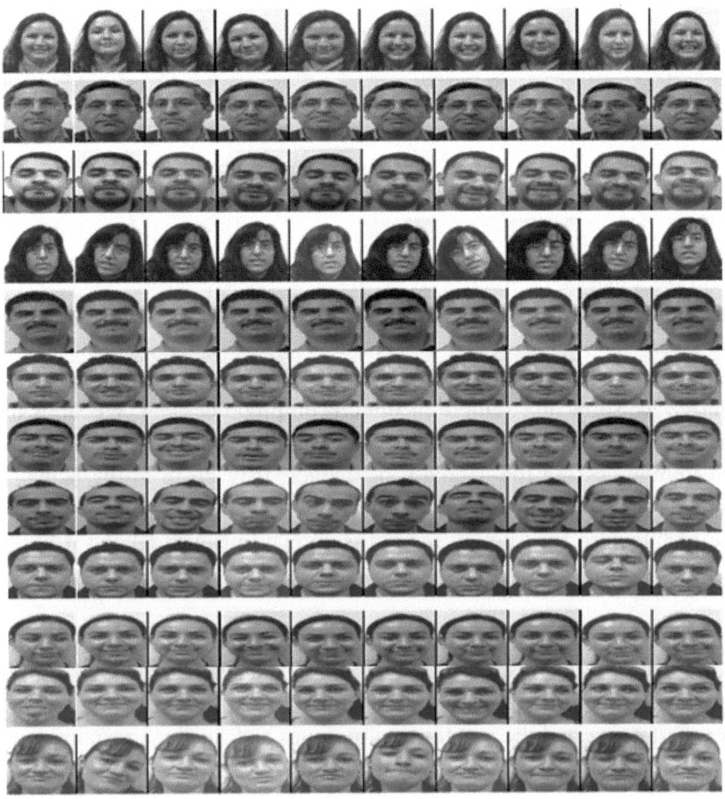

Fig. 11. Database of faces.

Fig. 12. ORL Database example of faces.

4.1.4 Test for the ORL Database (Formerly 'The ORL Database of Faces')

This Database of Faces, (formerly 'The ORL Database of Faces'), contains a set of face images taken between April 1992 and April 1994 at the ORL lab. The database was used in the context of a face recognition project carried out in collaboration with the Speech, Vision and Robotics Group of the Cambridge University Engineering Department.

Data is given as a set of 400 images, 10 photos of each individual appearing as individual files 1.bpm, 2.bpm,..., located at 40 directory named s1, s2, 10.bpm, s40. Each file.bpm (bitmap) is composed of a 14 byte header and image in 256 levels of gray as an array of 92 x 112 pixels.

There are ten different images of each of the 40 distinct subjects. For some subjects, the images were taken at different times, varying the lighting, facial expressions (open / closed eyes, smiling / not smiling) and facial details (glasses / no glasses). All the images were taken against a dark homogeneous background with the subjects in an upright, frontal position (with tolerance for some side movement). The image database is shown in Figure 12.

4.1.5 Design of the Neural Network Structure for Face Recognition

The number of neurons is allocated using an empirical expression created by Renato Salinas[8] and can be explained as follows:

- Input neurons: 2 * (k+m), the activation function is a Tangent sigmoid.
- Hidden neurons: (k+m), the activation function is a Tangent sigmoid.
- Output neurons: the activation function is a Tangent logarithm sigmoid.

Where k is the number of individuals to train and m is the number of samples to train for each individual. In this case, these were 40 individuals for the database of ORL, and 30 for the database of the institute and 7 samples therefore, k = 40, m = 7 and k = 30, m = 7.

Taking into account the previous data architecture to train the modules with the following form. The architecture is shown in the Figure 13.

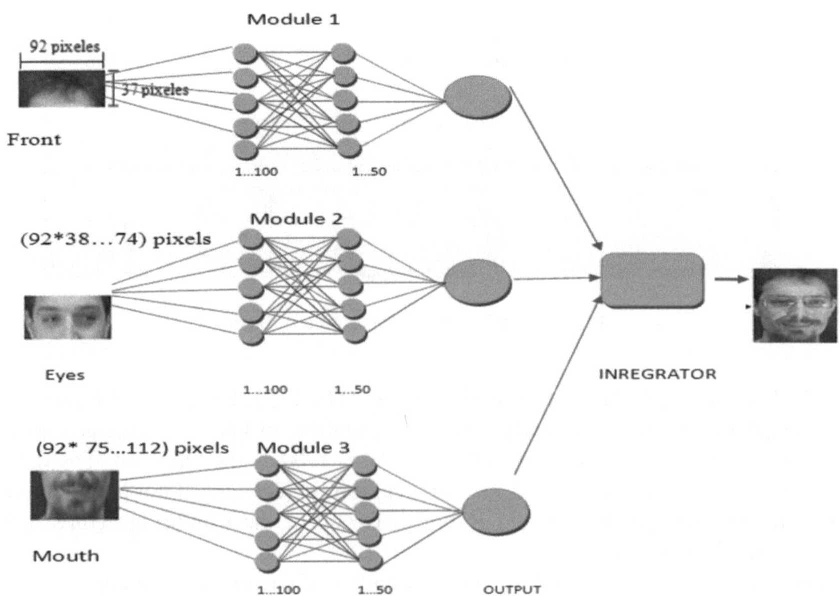

Fig. 13. Architecture of the neural network for face recognition.

5 Simulation Results

In this section, we only show the results obtained for each of the biometric measures taking the tests with 5 methods of training and to choose the one with the better results. In table 1 we show the different methods of training used with the data bases and the method for integration, which was the gating network.

We realize several trainings in order to arrive to these results, although these are on average the best results that we have obtained for each of the biometric measures.

Table 1. Training methods for the neural networks.

Abbreviation	Training method	Integrator
TRAINGDX	Gradient descendent with momentum and adaptive learning rate backpropagation	Gating Network
TRAINGDA	Gradient descendent with adaptive learning rate bagpropagation	Gating Network
TRAINSCG	Scaled conjugate gradient backpropagation	Gating Network
TRAINRP	Resilient backpropagation	Gating Network
TRAINCGB	Conjugate gradient backpropagation with Powell-Beales restarts	Gating Network

Table 2. Best simulation results without learning rate.

Training	Error Goal	Training Methods	Identification	Error	% Identification
1	0.0000001	Traingdx	117	3	117/120(97.50%)
2	0.0000001	Traingda	117	3	117/120(97.50%)
3	0.0000001	Trainscg	116	4	116/120(96.66%)
4	0.0000001	Trainrp	80	40	80/120(66.66%)
5	0.0000001	Traincgb	98	22	98/120(81.66%)

Table 3. Best simulation results with learning rate of 0.001.

Training	Error Goal	Training Methods	Identification	Error	% Identification
1	0.00000001	Traingdx	119	1	97/120(99.16%)
2	0.00000001	Traingda	117	3	97/120(97.50%)
3	0.00000001	Trainscg	116	4	95/120(96.66%)
4	0.00000001	Trainrp	80	40	56/120(66.66%)
5	0.00000001	Traincgb	98	22	83/120(81.66%)

5.1 Signature Results

The results for the measurement of biometric signatures were very good, we achieved an average recognition of 99%, which is considerably good for the development of the system, and these results are shown below in Tables 2 and 3.

Table 2 shows the best training without the use of learning rate, as can be seen; we have a percentage of recognition of 97.50 with the training methods traingdx and traingda.

Table 3 shows the best results that were obtained for signature with a learning rate of 0.001, increasing our percentage of identification here to a 99.16% and the best Method of training was traingdx.

Table 4. Best results without learning rate

Training	Error Goal	Training Methods	Identification	Error	% Identification
1	0.00000001	Traingdx	97	15	97/112(86.60%)
2	0.00000001	Traingda	97	15	97/112(86.60%)
3	0.00000001	Trainscg	95	17	95/112(84.82%)
4	0.00000001	Trainrp	56	56	56/112(50%)
5	0.00000001	Traincgb	83	29	83/112(74.10%)

Table 5. Best results with learning rate of 0.001

Training	Error Goal	Training Methods	Identification	Error	% Identification
1	0.0000001	Traingdx	99	13	92/112(88.39%)
2	0.0000001	Traingda	97	15	97/112(86.60%)
3	0.0000001	Trainscg	100	12	100/112(89.28%)
4	0.0000001	Trainrp	50	62	50/112(44.64%)
5	0.0000001	Traincgb	86	26	86/112(76.78%)

5.2 Fingerprint Results

Several tests were made to reach the results shown in tables 4 and 5. In Table 4 we show the results of tests made without learning rate and a target error of 0.00000001 of 86.60%.The best results were obtained with the methods of training traingdx and traingda.

Better results were obtained by adding a learning rate of 0.001 with the same goal error is achieved with this increase above the percentage of identification in a 89.28% as shown in table 5, the best training method was trainscg.

5.3 Face Results

For the face identification we have results for the data base of the institution and for the data base of the ORL, in both situations we performed different tests. In

Table 6. Results of original database take in the Tijuana Institute of Technology

Training	Error Goal	Training Methods	Epoch	Time	% Identification
1	0.0000001	Traingdx	500,500,500	2:11	60/60(100%)
2	0.0000001	Traingda	500,500,500	2:03	59/60(98.33%)
3	0.0000001	Trainscg	500,500,500	8:21	58/60(96.6%)
4	0.0000001	Trainrp	500,125,240	2:15	52/60(86.66%)
5	0.0000001	Traincgb	209,134,126	3:9	56/60(93.33%)

Table 7. Simulation results of cross validation

Train	Error Goal	Training Methods	Epoch	Time	% Identification
1	0.0000001	Traingdx	500,500,500	2:11	55/60(91.6%)
2	0.0000001	Traingda	500,500,500	2:03	54/60(90%)
3	0.0000001	Trainscg	500,500,500	8:21	57/60(95%)
4	0.0000001	Trainrp	500,125,240	2:15	52/60(86.66%)
5	0.0000001	Traincgb	209,134,126	3:9	56/60(93.33%)

table 6 we show results that include a case of 100% identification for the data base from the institution by means of the training method traingdx.

In order to verify the effectiveness of the result of the 100% of identification, tests of cross validation were realized where the positions of the images of the data base were alternated to verify that the percentage of identification persists, verifying if there is a result that fails as it is possible to be appreciated in table 7, another difference that can be appreciated with the cross validation is that the training method varies since with the normal data base we have the method of training traingdx and for these tests with which better results were achieved with trainscg, with a 95% of identification.

For the tests realized with the ORL database the results that have been obtained are shown in table 8. We show the results of the integration methods: Sugeno measures and Gating network.

For each of the biometric measures we have had very good results in an independent manner. It is important to consider the fact that in the future, when making the integration of the 3 biometric measures the percentage of identification will increase, because of the modularity and thus having a greater system reliability and performance.

5.3.1 Results of the ORL Database

Several training were done using different methods of training, changing the learning rate, to find a better result being the one shown in table 8, where greater identification has been found with the gating network integrator.

After having obtained the above mentioned results, three Cross-validations (permutations) were made to the original database, making 20 different trainings with each permutation and finally we obtain an average of three permutations. As with the original database were other methods of training can be used, like the gradient descent with momentum and adaptive learning rate back propagation (TRAINGDX), and the above-mentioned gradient Conjugate Escalade (trainscg), Conjugate gradient with Powell (traincgb).

Table 8. Results with training algorithms gradient Conjugate Escalade (trainscg), Conjugate gradient with Powell (traincgb), taking better performance the trainscg, both in training the neural network as in the identification results.

TRAINING METHOD	LEARNING RATE	GOAL ERROR	Trained images 70%	Images to Identify 30%	GATING NETWORK		Diffuse and measures integral of Sugeno	
					RECONOGNITION	IDENTIFY	RECONOGNITION	IDENTIFY
TRAINSCG 1	0.0001	0.001	280	120	280/280 100%	110/120 92%	280/280 100%	107/120 89%
	0.0001	0.001						
	0.0001	0.001						
TRAINCGB 2	0.000001	0.001	280	120	280/280 100%	113/120 94%	280/280 100%	114/120 91%
	0.0001	0.001						
	0.00001	0.001						
TRAINSCG	0.00001	0.001	280	120	280/280 100%	117/120 97%	280/280 100%	114/120 95%
	0.001	0.001						
	0.0001	0.001						

Table 9. The average of the three cross-validations (permutations) and the original database with gating network integration.

TRAINED	ERROR GOAL	TRAINING METHOD	LEARNING RATE	TOTAL TIME	IDENTIFICATION % SUGENO INTEGRAL
Data Base ORL Training 5	0.001	Trainscg	0.00001	00:03:40	95
	0.001	Trainscg	0.001		
	0.001	Trainscg	0.0001		
cross-validation1 Training 1	0.001	Traingdx	0.00001	00:04:50	95
	0.001	Traingdx	0.001		
	0.001	Traingdx	0.0001		
cross-validation2 Training 7	0.001	Trainscg	0.00001	00:05:01	92
	0.001	Traincgb	0.001		
	0.001	Trainscg	0.0001		
cross-validation3 Training 6	0.001	Trainscg	0.00001	00:02:56	94
	0.001	Traingdx	0.001		
	0.001	Trainscg	0.0001		
AVERAG					94

Table 10. The average of the three cross-validations (permutations) and the original database with Sugeno integration.

TRAINED	ERROR GOAL	TRAINING METHOD	LEARNING RATE	TOTAL TIME	IDENTIFICATION % Gating Network
Data Base ORL Training 7	0.001	Trainscg	0.00001	00:04:10	97
	0.001	Trainscg	0.001		
	0.001	Trainscg	0.0001		
cross-validation1 Training 1	0.001	Traingdx	0.00001	00:04:50	95
	0.001	Traingdx	0.001		
	0.001	Traingdx	0.0001		
cross-validation2 Training 6	0.001	Trainscg	0.00001	00:03:50	92
	0.001	Traingdx	0.001		
	0.001	Trainscg	0.0001		
cross-validation3 Training 6	0.001	Trainscg	0.00001	00:02:56	96
	0.001	Traingdx	0.001		
	0.001	Trainscg	0.0001		
AVERAGE					95

In table 9 we show the best trainings with cross-validation that were made to the ORL database, the first validation was with the best training method Traingdx with an error of 0.001 for the three modules and different learning rate, for the first module we used 0.00001, for the second module 0.001 and 0.0001 for the third module, with a total time of 4 minutes 10 seconds, achieving a 95% of identification with the Gating Network Integrator. The procedure is the same for the following three cross-validation, the results obtained in the following three were 97%, 92% and 96%respectively, calculating the average of them we obtained a 95% of identification.

In table 10 we show the best trainings with cross-validation that were made to the ORL database, the first validation was with the best training method Trainscg with an error of 0.001 for the three modules and different learning rate, for the first module we used 0.00001, for the second module 0.001 and 0.0001 for the third module, with a total time of 3 minutes 40 seconds, achieving a 95% of identification with the Sugeno integral. The procedure is the same for the following three cross-validation, the results obtained in the following three were 95%, 92% and 94% respectively, calculating the average of them we obtained a 94% of identification.

6 Conclusions

We can conclude that working with biometric measures is much more reliable than with other forms of authentication and the fact that combining several biometric measures helps to have a greater percentage of reliability because measures cover the deficiencies of the others within a system.

At this moment, we have good results with the independent biometric measures, so it is possible to say that good results may be obtained after having of the integration of 3 measures. We will try this in the near future.

Another thing to realize for the development of the system is as far as the face, for the data base of ORL is to develop an integrator with type-1 fuzzy logic and to test the operation and to compare results with the method that we have at this moment.

Acknowledgment

We would like to express our gratitude to CONACYT, and Tijuana Institute of Technology for the facilities and resources granted for the development of this research.

References

[1] Castillo, O., Melin, P.: Hybrid intelligent systems for time series prediction using neural networks, fuzzy logic, and fractal theory. IEEE Transactions on Neural Networks 13(6), 1395–1408 (2002)

[2] Faros, A.: Biologically Inspired Modular Neural Networks. Blacksburg, Virginia (May 2000); Kennedy, J., Eberhart, R.C.: Particle swarm optimization. In: Proceedings of IEEE International Conference on Neural Networks, Piscataway, NJ, pp. 1942–1948 (1995)

[3] Melin, P., Gonzalez, F., Martinez, G.: Pattern Recognition Using Modular Neural Networks and Genetic Algorithms. In: Proc. IC-AI 2004, Las Vegas, Nevada, USA, pp. 77–83 (2004)

[4] Romero, M.: Sistemas Modulares. Mezcla de expertos y sistemas Hibridos, Marzo (2004)

[5] Nuñez, R.: ITT, Redes Neuronales Modulares en Paralelo para el Reconocimiento de Personas (Febrero 2008)

[6] Pérez, A.: Reconocimiento y verificación de firmas manuscritas off-line basado en el seguimiento de sus trazos componentes (2002/2003), `http://eslova.com`

[7] Fu, H.-C., Lee, Y.-P., Chiang, C.-C., Hsiao-Tien: Divide and conquer learning and modular perceptron networks. In: Goldberg, D. (ed.) IEEE transaction on neural networks. Genetic Algorithms, vol. 12. Addison Wesley, Reading (1998)

[8] Salinas, R.: Departamento de Ingeniería Eléctrica. Universidad de Santiago de Chile, Red Neuronal de Arquitectura Paramétrica en Reconocimiento de Rostros, Universidad Diego Portales (agosto 2008), `http://cabierta.uchile.cl/revista/17/articulos/pdf/paper4.pdf`

[9] Romero, L.: Dpto. de Informática y Automática. Universidad de Salamanca- España. Calonge Cano Teodoro, Dpto de Informática. Universidad de Valladolid. España, Redes Neuronales y Reconocimiento de Patrones, `http://lisisu02.fis.usal.es/~airene/capit1.pdf`

[10] Sugeno, M.: Theory of Fuzzy integrals and its application. Tokyo Institute of Technology (1974)

Optimization of Modular Neural Networks with Interval Type-2 Fuzzy Logic Integration Using an Evolutionary Method with Application to Multimodal Biometry

Denisse Hidalgo[1], Patricia Melin[2], and Guillermo Licea[1]

[1] UABC University, Tijuana, México
[2] Tijuana Institute of Technology, Tijuana México
paulette1019@hotmail.com, epmelin@hafsamx.org, glicea@uabc.mx

Abstract. In this paper we describe a new evolutionary method to perform the optimization of a modular neural network applied to the case of multimodal biometry. Integration of responses in the modular neural network is performed using type-1 and type-2 fuzzy inference systems.

The proposed evolutionary method produces the best architecture of the modular neural network.

1 Introduction

Biometry is a discipline that studies the recognition of people through its physiological characteristics (fingerprint, face, retina, etc.) or of behavior (voice, signature, etc.). The interest to use biometric patterns to identify or verify the authenticity of the people has had a drastic increase that it is reflected in the appearance of diverse practical applications like identity passports that include biometrics characteristics.

Biometry can provide a true identification of people, since this technology is based on the recognition of unique corporal characteristics, reason why it recognizes the people based on who they are. Only biometric identification can provide a really efficient and precise control of the people, since it is possible to know with a high degree of certainty that the person that went through this form of recognition is the recognized person.

Multimodal biometry is based on using several biometric characteristics to improve the recognition rate and the reliability of the final recognition result. For this reason, in this paper, three biometric characteristics of a person are used to achieve a good recognition rate of humans [25].

The main goal of this research is develop the evolutionary method of complete optimization of the modular neural network application in multimodal biometrics. In this paper we describe the evolutionary method to develop and present the simulation results.

P. Melin et al. (Eds.): Bio-inspired Hybrid Intelligence Systems, SCI 256, pp. 111–121.
springerlink.com

This paper is organized as follows:

Section 2 shows an introduction to the theory of genetic algorithms, the section 3 describe the development of the evolutionary method; in the section 4 show the simulation results, section 5 shows the conclusions and finally references.

2 Genetic Algorithms

The GA is based on the Darwinian principle of natural selection for which the noted English philosopher, Herbert Spencer coined the phrase "Survival of the fittest". As a numerical optimizer, the solutions obtained by the GA are not mathematically oriented. Instead, GA possesses an intrinsic flexibility and the freedom to choose desirable optima according to design specifications. Whether the criteria of concern are nonlinear, constrained, discrete, multimodal, or NP hard, the GA is entirely equal to the challenge. In fact, because of the uniqueness of the evolutionary process and the gene structure of a chromosome, the GA processing mechanism can take the form of parallelism and multiobjective. These provide an extra dimension for solutions where other techniques may have failed completely.

The basic principles of the GA were first proposed by Holland. GA is inspired by the mechanism of natural selection where stronger individuals are likely the winners in a competing environment. Here, GA uses a direct analogy of such natural evolution. Through the genetic evolution method, an optimal solution can be found and represented by the final winner of the genetic game.

GA presumes that the potential solution of any problem is an individual and can be represented by a set of parameters. These parameters are regarded as the genes of a chromosome and can be structured by a string of values in binary form. A positive value, generally known as a fitness value, is used to reflect the degree of "goodness" of the chromosome for the problem which would be highly related with its objective value.

Throughout a genetic evolution, the fitter chromosome has a tendency to yield good quality offspring which means a better solution to any problem. In a practical GA application, a population pool of chromosomes has to be installed and these can be randomly set initially. The size of this population varies from one problem. In each cycle of genetic operation, termed as an evolving process, a subsequent generation is created from the chromosomes in the current population. This can only succeed if a group of these chromosomes, generally called "parents" or a collection term "mating pool" is selected via a specific selection routine. The genes of the parents are mixed and recombined for the production of offspring in the next generation. It is expected that from this process of evolution (manipulation of genes), the "better" chromosome will create a larger number of offspring, and thus has a higher chance of surviving in the subsequent generation, emulating the survival-of-the-fittest mechanism in nature.

The cycle of evolution is repeated until a desired termination criterion is reached. This criterion can also be set by the number of evolution cycles (computational runs), or the amount of variation of individuals between different generations, or a pre-defined value of fitness. In order to facilitate the GA evolution cycle, two fundamental operators: Crossover and Mutation are required, although the selection routine can be termed as the other operator [1].

3 Evolutionary Method Description

Based on the theory described above, a new method for optimization of Modular Neural Networks with Type-2 Fuzzy Integration using an Evolutionary Method with application in Multimodal Biometry is proposed.

The goal of the research is the development of a general evolutionary approach to optimize modular neural networks including the module of response integration. In particular, the general method includes optimizing the type-2 fuzzy system that performs the integration of responses in the modular network, as well as the optimization of the complete architecture of the modular neural network, as number of modules, layers and neurons. The purpose of obtaining the optimal architecture is to obtain better recognition rates and improve the efficiency of the hybrid system of pattern recognition.

In figure 1we see the specific points to the general scheme of the evolutionary method.

We have a modular neural network with fuzzy integration, which is optimized with the proposed evolutionary approach.

First, the optimization of the complete architecture of the modular neural network, as number of modules, layers and neurons is performed. Later, the optimize the genetic algorithm of the method of integration, where this method of integration uses type-2 fuzzy inference systems. After obtaining the optimization of the architecture of the MNN and the algorithm of integration method, developed the new method of Optimization using General Evolutionary Method with Application

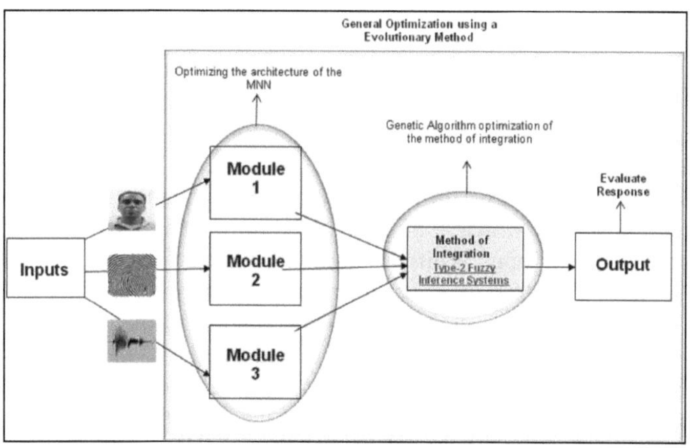

Fig. 1. General scheme of evolutionary method.

in Multimodal Biometry; will be implemented on Evolutionary Method developed in the Generalized Optimization of Modular Neural Networks with Application in Multimodal Biometry with type-2 fuzzy integration, and then pass to the validation of results and conclusions to compare statistically the method.

4 Simulation Results

Previously we have worked on a comparative study of type-1 and type-2 fuzzy integration for modular neural networks in multimodal biometry, optimized by genetic algorithms. We used different chromosomes structures. In the figure 2 we show the General Scheme of the pattern recognition system. The input data used in the modular architecture for pattern recognition as given in [8, 10].

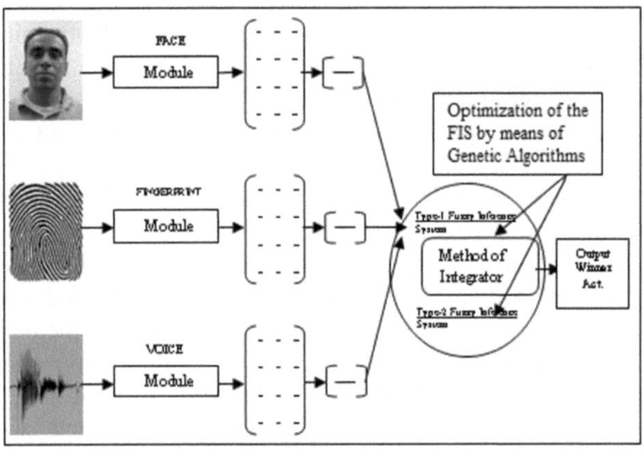

Fig. 2. General Scheme of the pattern recognition system

The fuzzy systems were of Mamdani type with three inputs and one output, three triangular, trapezoidal or Gaussians type-1 or type-2 membership functions; therefore, we obtained various systems integrators as fuzzy inference to test the modular neural networks.

To obtain a type-1 FIS with triangular membership functions we used a chromosome of 36-bits, of which 9 bits were used for the parameters of the first input, 9 bits for the second input, 9 bits for the third input and finally 9 bits for the output. To obtain FIS with trapezoidal membership functions used a chromosome of 48-bits, of which 12 bits were used to the parameters of the first input, 12 bits for the second input, 12 bits for the third input and finally 12 bits for the output. To obtain FIS with Gaussian membership functions used a chromosome of 24-bits, of which 6 bits were used to the parameters of the first input, 6 bits for the second input, 6 bits for the third input and finally 6 bits for the output. This is shown in Figure 3.

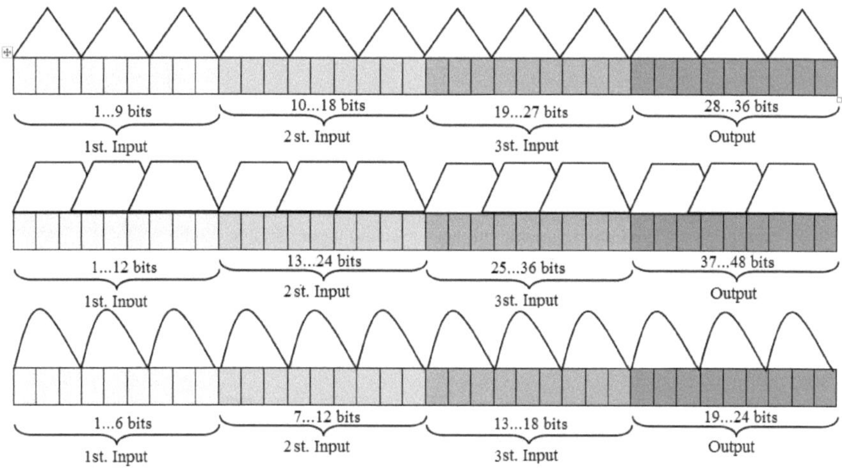

Fig. 3. Chromosomes used to type-1 FIS.

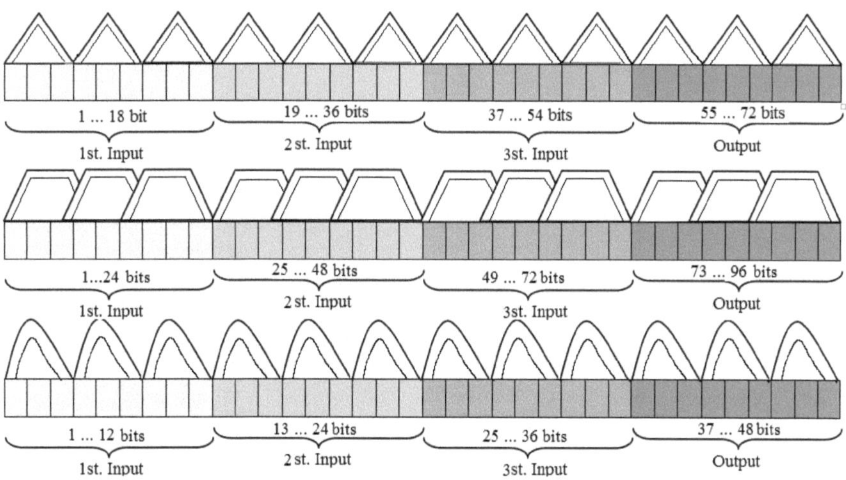

Fig. 4. Chromosomes used to type-2 FIS.

To obtain type-2 FIS with triangular membership functions used a chromosome of 72-bits, of which 18 bits were used to the parameters of the first input, 18 bits for the second input, 18 bits for the third input and finally 18 bits for the output. To obtain FIS with trapezoidal membership functions used a chromosome of 96-bits, of which 24 bits were used to the parameters of the first input, 24 bits for the second input, 24 bits for the third input and finally 24 bits for the output. To obtain FIS with Gaussians membership functions used a chromosome of 48-bits, of which 12 bits were used to the parameters of the first input, 12 bits for the second input, 12 bits for the third input and finally 12 bits for the output. This is shown in Figure 4 [2].

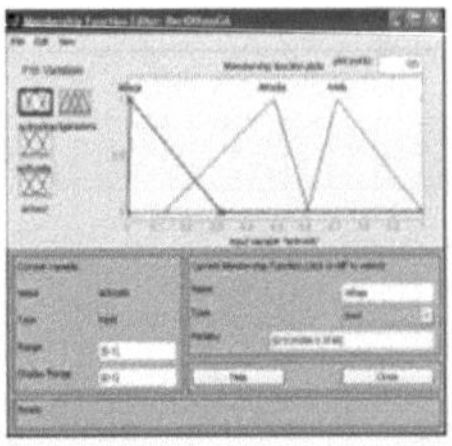
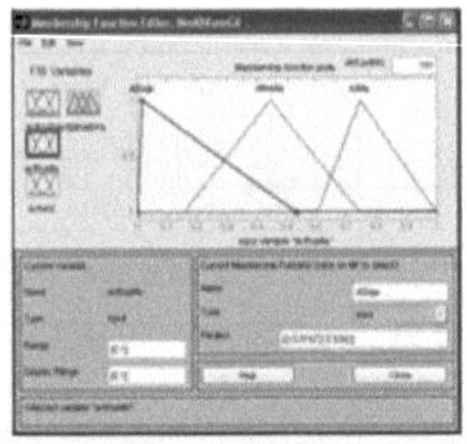

First input variable (higher activation of the face) | Second input variable (higher activation of the fingerprint)

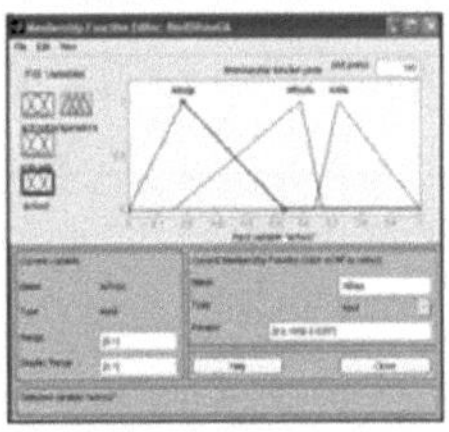
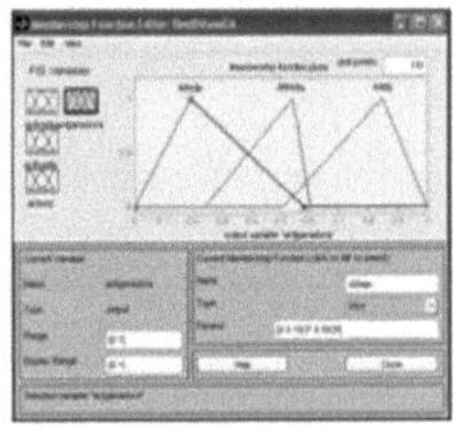

Third input variable (higher activation of the voice) | Variable of output (winner activation)

Fig. 5. The best Type-1 FIS with triangular membership function.

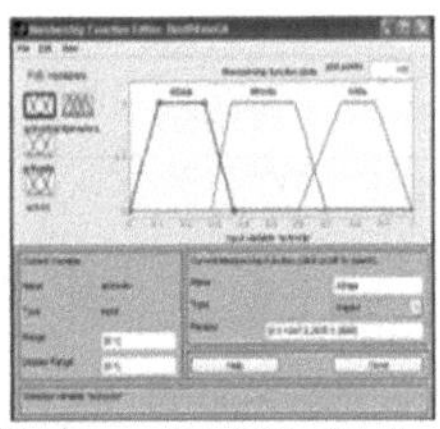
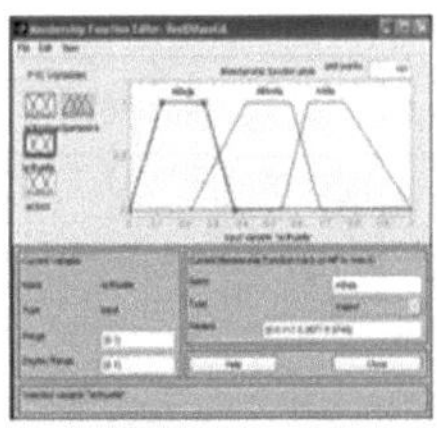

First input variable (higher activation of the face) | Second input variable (higher activation of the fingerprint)

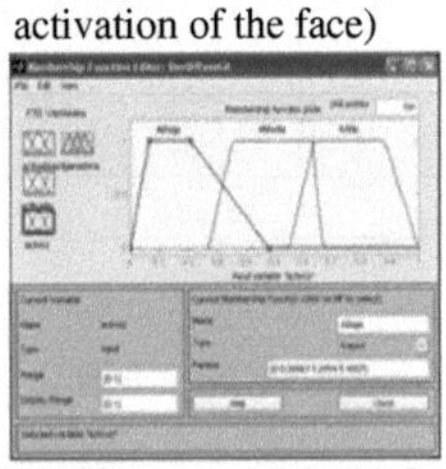
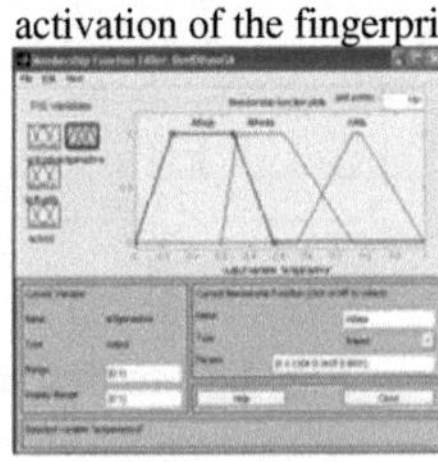

Third input variable (higher activation of the voice) | Variable of output (winner activation)

Fig. 6. The best Type-1 FIS with trapezoidal membership function.

Type-1 and Type-2 Fuzzy Integrations

In the following section we show the best Fuzzy Inference Systems (FIS), which was obtained with the optimization of the Genetic Algorithms. In the figure 5, 6 and 7 we show the best type-1 FIS with membership functions of triangular, trapezoidal and Gaussians respectively.

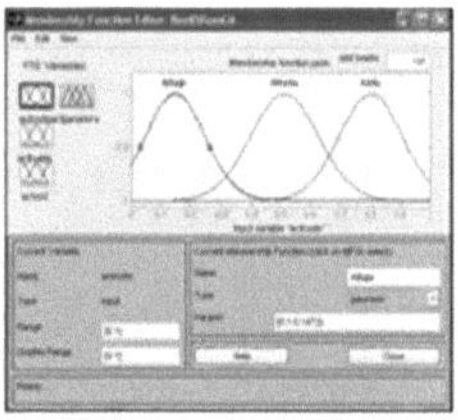

First input variable (higher activation of the face)

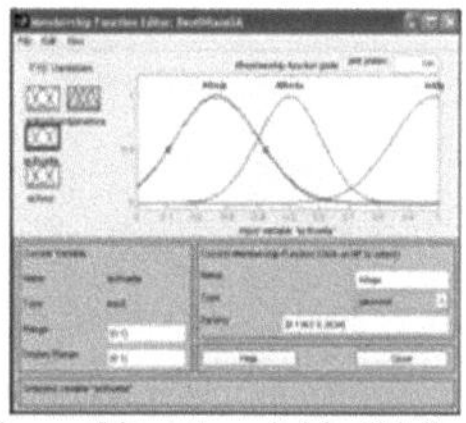

Second input variable (higher activation of the fingerprint)

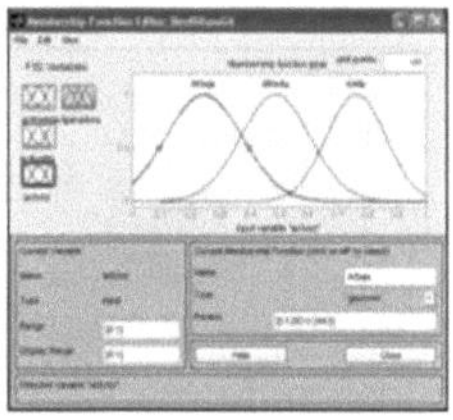

Third input variable (higher activation of the voice)

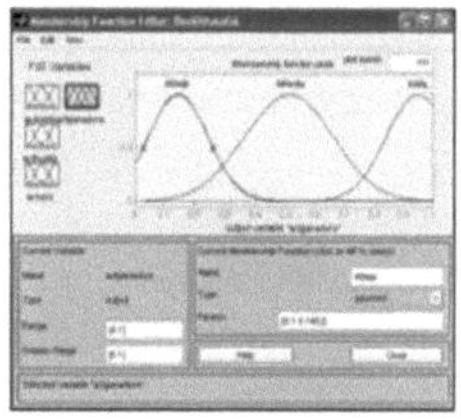

Variable of output (winner activation)

Fig. 7. The best Type-1 FIS with Gaussian membership function.

First input variable (higher activation of the face)

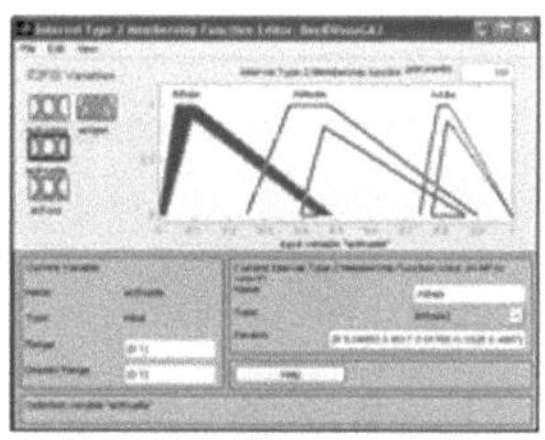

Second input variable (higher activation of the fingerprint)

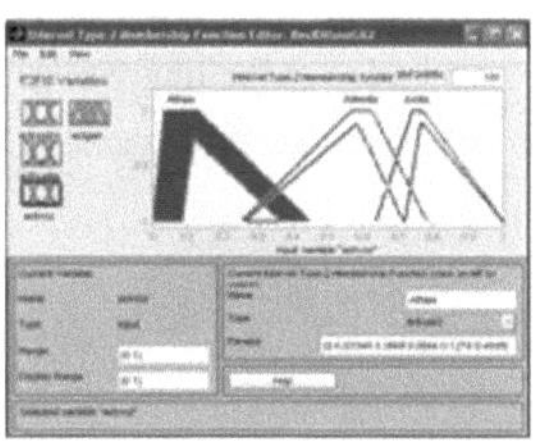

Third input variable (higher activation of the voice)

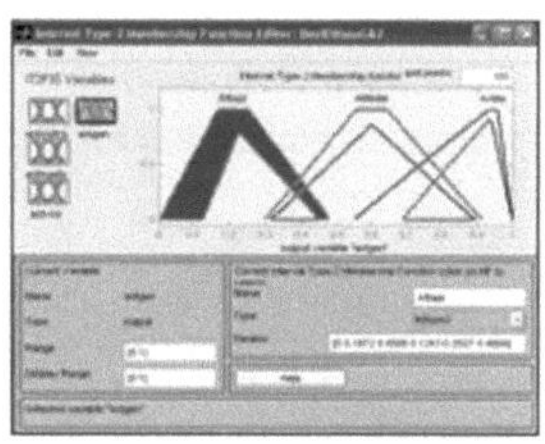

Variable of output (winner activation)

Fig. 8. The best Type-2 FIS with Triangular membership function.

Now, in the figure 8, 9 and 10 we show the best Type-2 FIS with membership functions of triangular, trapezoidal and Gaussians respectively; which were obtained whit the optimization of Genetics Algorithms.

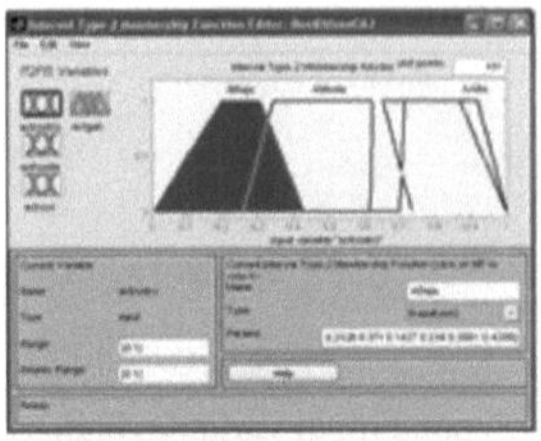

First input variable (higher activation of the face)

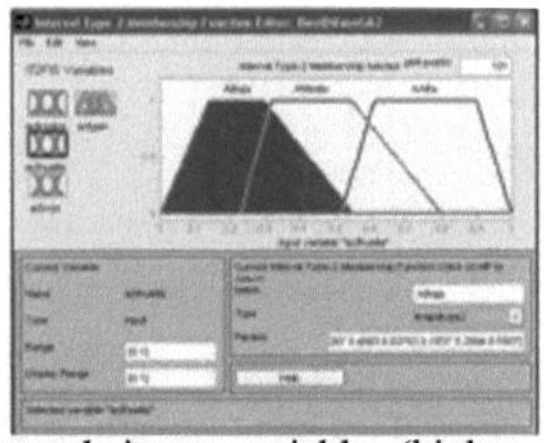

Second input variable (higher activation of the fingerprint)

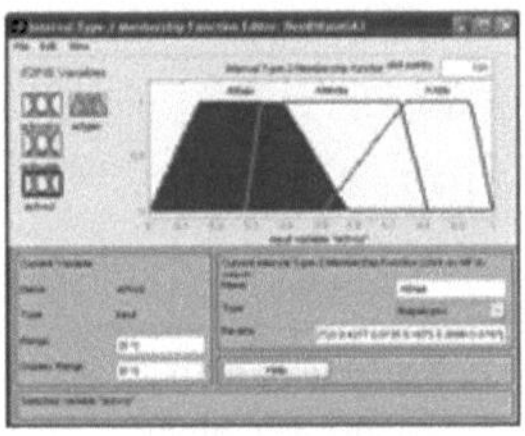

Third input variable (higher activation of the voice)

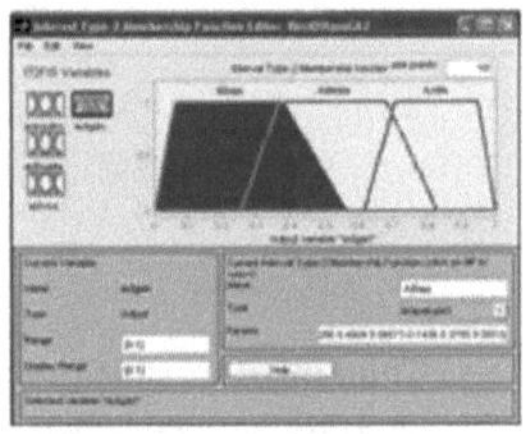

Variable of output (winner activation)

Fig. 9. The best Type-2 FIS with Trapezoidal membership function.

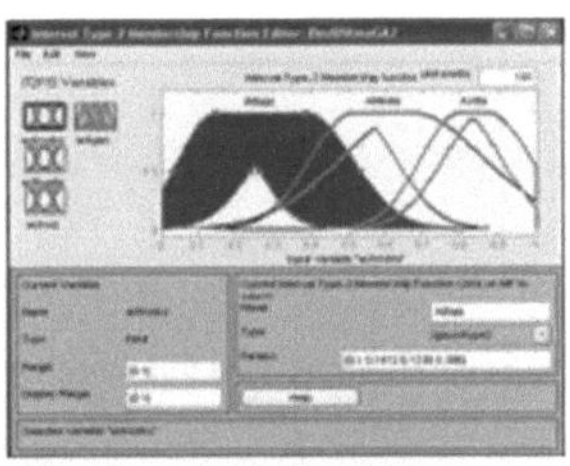

First input variable (higher activation of the face)

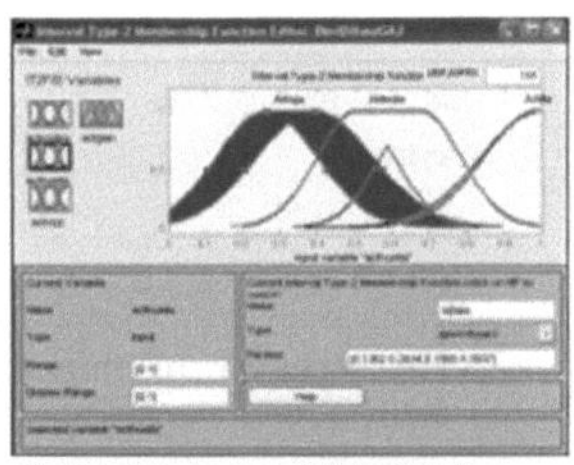

Second input variable (higher activation of the fingerprint)

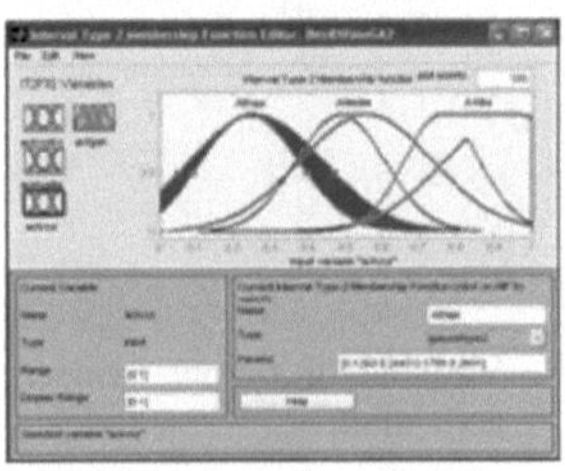

Third input variable (higher activation of the voice)

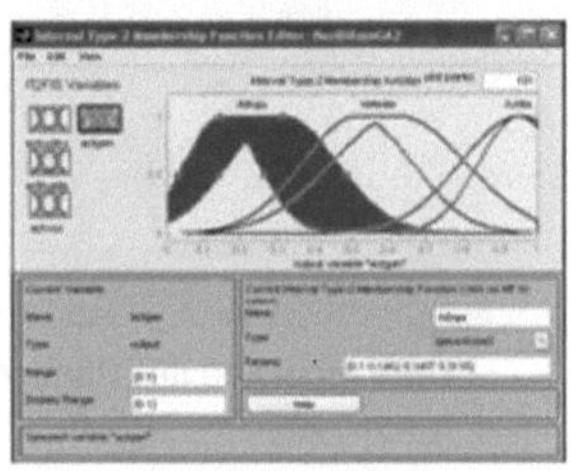

Variable of output (winner activation)

Fig. 10. The best Type-2 FIS with Gaussian membership function.

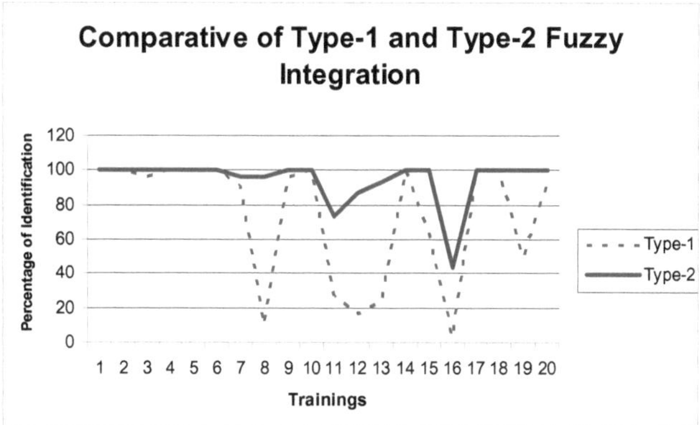

Fig. 11. Comparison of Integration with type-1 and type-2 Fuzzy Systems

Comparative Integration with Type-1 and Type-2 Fuzzy Inference Systems
After the modular neural network trainings were obtained, we make the integration of the modules with the type-1 and type-2 optimized fuzzy systems. Next we show the type-1 and type-2 graphics and table with the 20 modular neural network trainings and the percentage of the identification (See Figure 11). We can appreciate that type-2 Fuzzy Logic is better [2].

Table 1 shows the average percentage of recognition of the type-1 and type-2 Fuzzy Inference Systems that we tested on the last experiment.

Table 1. Comparative table of average percentage integration with type-1 and type-2 fuzzy inference systems

Type-1	Type-2
Average percentage of identification	Average percentage of identification
73.50 %	94.50 %

5 Conclusions

In this paper we presented a comparison study between type-1 and type-2 fuzzy systems as integration methods of modular neural networks. The comparison was made using different simulations of modular neural networks trained with the faces, fingerprints and voices of a database of persons for recognition. Simulations results of the modular neural networks with fuzzy systems as integration modules were good. Type-2 Fuzzy Systems are shown to be a superior method for integration of responses in Modular Neural Networks application in multimodal biometrics. Future work we will continue with the optimization of the architecture of the modular neural network.

References

[1] Man, K.F., Tang, K.S., Kwong, S.: Genetic Algorithms, Concepts and Designs. Springer, Heidelberg (1999)

[2] Hidalgo, D., Melin, P., Castillo, O.: Type-1 and Type-2 Fuzzy Inference Systems as Integration Methods in Modular Neural Networks for Multimodal Biometry and its Optimization with Genetic Algorithms. Journal of Automation, Mobile Robotics & Intelligent Systems 2(1) (2008) ISSN 1897-8649

[3] Jang, J.-S.R., Sun, C.-T., Mizutani, E.: Neuro-Fuzzy and Soft Computing. A Computational Approach to Learning and Machine. Intelligence Prentice Hall (1997)

[4] Castro, J.R.: Tutorial Type-2 Fuzzy Logic: theory and applications. Universidad Autónoma de Baja California-Instituto Tecnológico de Tijuana (October 9, 2006), `http://www.hafsamx.org/cis-chmexico/seminar06/tutorial.pdf`

[5] Melin, P., Castillo, O., Gómez, E., Kacprzyk, J., Pedrycz, W.: Analysis and Design of Intelligent Systems Using Soft Computing Techniques. In: Advances in Soft Computing, vol. 41. Springer, Heidelberg (2007)

[6] The 2007 International Joint Conference on Neural Networks, IJCNN, Conference Proceedings, Orlando, Florida, USA, August 12-17. IEEE, Los Alamitos (2007) IEEE Catalog Number: 07CH37922C; ISBN: 1-4244-1380-X, ISSN: 1098-7576, ©2007

[7] Melin, P., Castillo, O.: Hybrid Intelligent Systems for Pattern Recognition Using Soft Computing: An Evolutionary Approach for Neural Networks and Fuzzy Systems. Studies in Fuzziness and Soft Computing (2005) (Hardcover - April 29, 2005)

[8] Alvarado-Verdugo, J.M.: Reconocimiento de la persona por medio de su rostro y huella utilizando redes neuronales modulares y la transformada wavelet. Instituto Tecnológico de Tijuana (2006)

[9] Melin, P., Castillo, O., Kacprzyk, J., Pedrycz, W.: Hybrid Intelligent Systems. Studies in Fuzziness and Soft Computing (Hardcover - December 20, 2006)

[10] Ramos-Gaxiola, J.: Redes Neuronales Aplicadas a la Identificación de Locutor Mediante Voz Utilizando Extracción de Características, Instituto Tecnológico de Tijuana (2006)

[11] Mendoza, O., Melin, P., Castillo, O., Licea, P.: Type-2 Fuzzy Logic for Improving Training Data and Response Integration in Modular Neural Networks for Image Recognition. In: Melin, P., Castillo, O., Aguilar, L.T., Kacprzyk, J., Pedrycz, W. (eds.) IFSA 2007. LNCS (LNAI), vol. 4529, pp. 604–612. Springer, Heidelberg (2007)

[12] Urias, J., Melin, P., Castillo, O.: A Method for Response Integration in Modular Neural Networks using Interval Type-2 Fuzzy Logic. In: FUZZ-IEEE 2007, Number 1 in FUZZ, London, UK, July 2007, pp. 247–252. IEEE, Los Alamitos (2007)

[13] Urias, J., Hidalgo, D., Melin, P., Castillo, O.: A Method for Response Integration in Modular Neural Networks with Type-2 Fuzzy Logic for Biometric Systems. In: Melin, P., et al. (eds.) Analysis and Design of Intelligent Systems using Soft Computing Techniques, 1st edn. Studies in Fuzziness and Soft Computing, vol. 1(1), pp. 5–15. Springer, Heidelberg (2007)

[14] Urias, J., Hidalgo, D., Melin, P., Castillo, O.: A New Method for Response Integration in Modular Neural Networks Using Type-2 Fuzzy Logic for Biometric Systems. In: Proc. IJCNN-IEEE 2007, Orlando, USA, August 2007. IEEE, Los Alamitos (2007)

[15] Melin, P., Castillo, O., Gómez, E., Kacprzyk, J.: Analysis and Design of Intelligent Systems using Soft Computing Techniques. Advances in Soft Computing (Hardcover - July 11, 2007)
[16] Mendoza, O., Melin, P., Castillo, O., Licea, P.: Modular Neural Networks and Type-2 Fuzzy Logic for Face Recognition. In: Reformat, M. (ed.) Proceedings of NAFIPS 2007, San Diego, June 2007, vol. (1), pages CD Rom. IEEE, Los Alamitos (2007)
[17] Zadeh, L.A.: Knowledge representation in Fuzzy Logic. IEEE Transactions on knowledge data engineering 1, 89 (1989)
[18] Zadeh, L.A.: Fuzzy Logic = Computing with Words. IEEE Transactions on Fuzzy Systems 4(2), 103 (May 1996)
[19] Mendel, J.M.: UNCERTAIN Rule-Based Fuzzy Logic Systems, Introduction and New Directions. Prentice Hall, Englewood Cliffs (2001)
[20] Mendel, J.M.: Why We Need Type-2 Fuzzy Logic Systems? Article is provided courtesy of Prentice Hall, May 11 (2001), http://www.informit.com/articles/article.asp?p=21312&rl=1
[21] Mendel, J.M.: Uncertainty: General Discussions, Article is provided courtesy of Prentice Hall, By Jerry Mendel, May 11 (2001), http://www.informit.com/articles/article.asp?p=21313
[22] Mendel, J.M., Bob-John, R.I.: Type-2 Fuzzy Sets Made Simple. IEEE Transactions on Fuzzy Systems 10(2), 117 (2002)
[23] Karnik, N., Mendel, J.M.: Operations on type-2 fuzzy sets, Signal and Image Processing Institute, Department of Electrical Engineering-Systems. University of Southern California, Los Angeles, USA, May 11 (2000)
[24] Zadeh, L.A.: Fuzzy Logic. Computer 1(4), 83–93 (1998)
[25] Hidalgo, D., Castillo, O., Melin, P.: Interval type-2 fuzzy inference systems as integration methods in modular neural networks for multimodal biometry and its optimization with genetic algorithms. International Journal of Biometrics 1(1), 114–128 (2008)

Comparative Study of Fuzzy Methods for Response Integration in Ensemble Neural Networks for Pattern Recognition

Miguel Lopez, Patricia Melin, and Oscar Castillo

Computer Science in the Graduate Division Tijuana,
Institute of Technology Tijuana, B. C., Mexico
pmelin@tectijuana.edu.mx, ocastillo@tectijuana.edu.mx

Abstract. We describe in this paper a new method for response integration in ensemble neural networks with Type-1 Fuzzy Logic and Type-2 Fuzzy Logic using Genetic Algorithms (GA's) for optimization. In this paper we consider pattern recognition with ensemble neural networks for the case of fingerprints to the test proposed method of response integration. An ensemble neural network of three modules is used. Each module is a local expert on person recognition based on their biometric measure (Pattern recognition for fingerprints). The Response Integration method of the ensemble neural networks has the goal of combining the responses of the modules to improve the recognition rate of the individual modules. First we use GA's to optimize the fuzzy rules of The Type-1 Fuzzy System and Type-2 Fuzzy System to test the proposed method of response integration and after using GA's to optimize the membership function of The Type-1 Fuzzy Logic and Type-2 Fuzzy logic to test the proposed method of response integration and finally show the comparison of the results between these methods. We show in this paper a comparative study of fuzzy methods for response integration and the optimization of the results of a type-2 approach for response integration that improves performance over the type-1 logic approaches.

1 Introduction

In single neural network methods, the neural network learning problem is often formulated as an optimization problem, i.e., minimizing certain criteria, e.g., minimum error, fastest learning, lowest complexity, etc., about architectures. Learning algorithms, such as backpropagation (BP) (Rumelhart, Hinton & Williams, 1986), are used as optimization algorithms to minimize an error function. Despite the different error functions used, these learning algorithms reduce a learning problem to the same kind of optimization problem [30].

Learning is different from optimization because we want the learned system to have the best generalization, which is different from minimizing an error function. The neural network with the minimum error on the training set does not necessarily

P. Melin et al. (Eds.): Bio-inspired Hybrid Intelligence Systems, SCI 256, pp. 123–140.
springerlink.com

have the best generalization unless there is equivalence between generalization and the error function. Unfortunately, measuring generalization exactly and accurately is almost impossible in practice (Wolpert, 1990), although there are many theories and criteria on generalization, such as the minimum description length (Rissanen, 1978), Akaike's information criteria (Akaike, 1974) and minimum message length (Wallace & Patrick, 1991). In practice, these criteria are often used to define better error functions in the hope that minimizing the functions will maximize generalization.

While better error functions often lead to better generalization of learned systems, there is no guarantee. Regardless of the error functions used, single network methods are still used as optimization algorithms. They just optimize different error functions. The nature of the problem is unchanged. While there is little we can do in single neural network methods, there are opportunities in neural network ensemble methods. Neural network ensembles adopt the divide-and-conquer strategy. Instead of using a single network to solve a task, a neural network ensemble combines a set of neural networks which learn to subdivide the task and thereby solve it more efficiently and elegantly. A neural network ensemble offers several advantages over a monolithic neural network.

First, it can perform more complex tasks than any of its components (i.e., individual neural networks in the ensemble). Secondly, it can make an overall system easier to understand and modify. Finally, it is more robust than a monolithic neural network and can show graceful performance degradation in situations where only a subset of neural networks in the ensemble are performing correctly. Given the advantages of neural network ensembles and the complexity of the problems that are beginning to be investigated, it is clear that the neural network ensemble method will be an important and pervasive problem-solving technique.

The idea of designing an ensemble learning system consisting of many subsystems can be traced back to as early as 1958 (Selfridge, 1958; Nilsson, 1965). Since the early 1990s, algorithms based on similar ideas have been developed in many different but related forms, such as neural network ensembles (Hansen & Salamon, 1990; Sharkey, 1996), mixtures of experts (Jacobs, Jordan, Nowlan & Hinton, 1991; Jacobs & Jordan, 1991; Jacobs, Jordan & Barto, 1991; Jacobs, 1997), various boosting and bagging methods (Drucker, Cortes, Jackel, LeCun & Vapnik, 1994; Schapire, 1990; Drucker, Schapire & Simard, 1993) and many others. There are a number of methods of designing neural network ensembles. To summarize, there are three ways of designing neural network ensembles in these methods: independent training, sequential training and simultaneous training.

In order to use the feedback from the integration, simultaneous training methods train a set of networks together. Negative correlation learning (Liu & Yao, 1998a, 1998b, 1999) and the mixtures-of-experts (ME) architectures (Jacobs et al., 1991; Jordan & Jacobs, 1994) are two examples of simultaneous training methods. The idea of negative correlation learning is to encourage different individual networks in the ensemble to learn different parts or aspects of the training data, so that the ensemble can better learn the entire training data. In negative correlation learning, the individual networks are trained simultaneously rather than independently or sequentially. This provides an opportunity for the individual networks to interact with each other and to specialize. That is, the lower mutual information two variables have, the more different they are. By minimizing the mutual information between variables extracted by two neural networks, they are forced to convey different information about some features of their input. The idea of minimizing mutual information is to encourage different individual networks to learn different parts or aspects of the training data so that the ensemble can learn the whole training data better.

Based on the decision boundaries and correct response sets, negative correlation learning has been further studied on two pattern classification problems. The purpose of examining the decision boundaries and the correct response sets is not only to illustrate the learning behavior of negative correlation learning, but also to cast light on how to design more effective neural network ensembles. The experimental results showed the decision boundary of the trained neural network ensemble by negative correlation learning is almost as good as the optimum decision boundary.

2 Biometric Systems

The earliest form of Biometrics appeared on the scene back in the 1800's. Alphonse Bertillon, a Perisian anthropologist and police desk clerk, developed a method for identifying criminals that became known as Bertillonage. Bertillonage was a form of anthropometry, a system by which measurements of the body are taken for classification and comparison purposes. Bertillons system of anthropometry required numerous and percise measurements of the bony parts of a humans anatomy for identification. It also involved recording shapes of the body in relation to movements and differential markings on the surface of the body such as scars, birth marks, tattoos, etc. Bertillon estimated that the odds of duplicate records were 286,435,456 to 1 if 14 traits were used. This was the primary system of criminal identification used during the 19th century [29].

Bertillons system of identification was not without fault. For example, it relied heavily on precise measurements for identification purposes, and yet two people working on measurements for the same person would record different findings. The measurements taken were also only thought to be unique and accurate in adulthood. Therefore, someone who committed a crime prior to adulthood would not have their measurements on record. Additionally, it turned out to be the case that the features by which Bertillon based his identification system were not unique to any one individual. This led to the possibility of one person being convicted of another persons crimes. This possibility became abundantly clear in 1903 when a Will West was confused with a William West. Though it would later turn out to be the case that the two were identical twins, the issues posed by the Bertillonage system of identification were clear.

Because of the amount of time and effort that went in to painstakingly collecting measurements and the overall inaccuracy of the process, Bertillonage was quickly replaced when fingerprinting emerged on the scene as a more efficient and accurate means of identification. Fingerprinting, as a means of identification, proved to be infallable. It was accepted that each individual possessed a uniquely identifiable and unchanging fingerprint. This new system of identification was accepted as more reliable than Bertillonage.

Fingerprinting can be traced as far back as the 14th century in China. Though the use was most likely as a signature and the unique identification abilities of the fingerprint not entirely known. Fingerprints were first looked at as a form of criminal identification by Dr. Henry Faulds who noticed fingerprints on ancient pottery while working in Tokyo.

He first published his ideas about using fingerprints as a means of identifying criminals in the scientific journal, *Nature* in 1880. William Herschel, while working in colonial India, also recognized the unique qualities that fingerprints had to offer as a means of identification in the late 1870's. He first began using fingerprints as a form of signature on contracts with locals. Sir Francis Galton, who had been privy to Faulds research through his uncle, Charles Darwin, would also be credited as making significant advancement to fingerprint identification. Galton ascertained that no two fingerprints were alike, not even on a set of identical twins. He noted that differentiating characteristics could be best observed in the ridge of a fingerprint and that this fingerprint would remain reliable and unchanging and could be used for identification throughout an individual's life. However, it had never been officially recognized as to which of these three men was the first to discover fingerprinting as a means of identification. The Henry Classification system, named after Edward Henry who developed and first implemented the system in 1897 in India, was the first method of classification for fingerprint identification based on physiological characteristics. The system assigns each individual finger a numerical value (starting with the right thumb and ending with the left pinky) and divides fingerprint records into groupings based on pattern types.

The system makes it possible to search large numbers of fingerprint records by classifying the prints according to whether they have an "arch," "whorl," or "loop" and the subsequently assigned numerical value. In 1901 the Henry system was introduced in England. In 1902 the New York Civil service began testing the Henry method of fingerprinting with the Army, Navy, and Marines all adopting the method by 1907. From this point on, the Henry System of fingerprinting became the system most commonly used in English speaking countries.

2.1 Fingerprint Identification Systems

Fingerprint Identification is the method of identification using the impressions made by the minute ridge formations or patterns found on the fingertips. No two persons have exactly the same arrangement of ridge patterns, and the patterns of any one individual remain unchanged throughout life. Fingerprints offer an infallible means of personal identification. Other personal characteristics may change, but fingerprints do not.

Fingerprints can be recorded on a standard fingerprint card or can be recorded digitally and transmitted electronically to the FBI for comparison. By comparing fingerprints at the scene of a crime with the fingerprint record of suspected

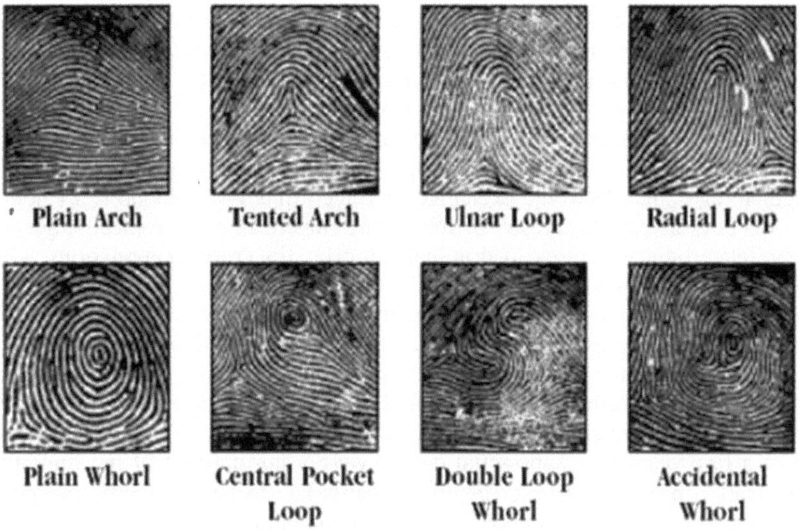

Fig. 1. Samples of the kinds fingerprints.

persons, officials can establish absolute proof of the presence or identity of a person, in the figure 1 shown something samples of the kinds fingerprints.

The first year for the first known systematic use of fingerprint identification began in the United States is 1902. The New York Civil Service Commission established the practice of fingerprinting applicants to prevent them from having better qualified persons take their tests for them. The New York state prison system began to use fingerprints for the identification of criminals in 1903. In 1904 the fingerprint system accelerated when the United States Penitentiary at Leavenworth, Kansas, and the St. Louis, Missouri, Police Department both established fingerprint bureaus. During the first quarter of the 20th century, more and more local police identification bureaus established fingerprint systems. The growing need and demand by police officials for a national repository and clearinghouse for fingerprint records led to an Act of Congress on July 1, 1921, establishing the Identification Division of the FBI.

In 1924 the Identification Division of the Federal Bureau of Investigation (FBI) was established to provide one central repository of fingerprints. When the Identification Division was established its purpose was to provide a central repository of criminal identification data for law enforcement agencies throughout the Nation. However, in 1933 the United States Civil Service Commission (now known as the Office of Personnel Management) turned the fingerprints of more that 140, 000 Government employees and applicants over to the FBI. Therefore, a Civil Identification Section was established. These innovations marked the initiation of the FBI's Civil File which was destined to dwarf the criminal files in size. In 1992 the Identification Division was re-established as the Criminal Justice Information Services Division (CJIS).

At the forefront of fingerprint biometric technology is the Integrated Automated Fingerprint Identification System (IAFIS). The Integrated Automated Fingerprint Identification System, more commonly known as IAFIS, is a national fingerprint and criminal history system maintained by the Federal Bureau of Investigation (FBI), Criminal Justice Information Services (CJIS) Division. The IAFIS provides automated fingerprint search capabilities, latent searching capability, electronic image storage, and electronic exchange of fingerprints and responses, 24 hours a day, 365 days a year. As a result of submitting fingerprints electronically, agencies receive electronic responses to criminal ten-print fingerprint submissions within two hours and within 24 hours for civil fingerprint submissions.

The IAFIS maintains the largest biometric database in the world, containing the fingerprints and corresponding criminal history information for more than 47 million subjects in the Criminal Master File. The fingerprints and corresponding criminal history information are submitted voluntarily by state, local, and federal law enforcement agencies.

Just a few years ago, substantial delays were a normal part of the fingerprint identification process, because fingerprint cards had to be physically transported and processed. A fingerprint check could often take three months to complete. The FBI formed a partnership with the law enforcement community to revitalize the fingerprint identification process, leading to the development of the IAFIS. The IAFIS became operational in July 1999.

In the last decade, interest in fingerprint-based biometric systems has grown significantly [27]. Activity on this topic increased in both academia and industry as several research groups and companies developed new algorithms and techniques for fingerprint recognition and as many new electronic fingerprint acquisition sensors were launched into the marketplace.

Nevertheless, before this initiative, only a few benchmarks have been available for comparing developments in this area and developers usually perform internal tests over self-collected databases. The lack of standards has unavoidably lead to the dissemination of confusing, incomparable and irreproducible results, sometimes embedded in research papers and sometimes enriching the commercial claims of marketing brochures. The aim of this initiative is to take the first steps toward the establishment of a common basis, both for academia and industry, to better understand the state-of-the-art and what can be expected from this technology in the future.

3 Ensemble Neural Network

This paper provides the main characteristics of the ensembles that will be further tested in the experiments. Correspondingly, we here describe the particular topology of the neural networks.

3.1 Multilayer Perceptron

Probably, the multilayer perceptron is one of the most popular connectionist models. It organizes the representation of knowledge in the hidden layers and has a very high power of generalization.

The typical topology consists of three sequential layers: input, hidden and output [15]. For the ensemble of multilayer perceptrons proposed in this study, all the networks have k input units and m output units corresponding to the number of input features (or attributes) and data classes, respectively. Also, each network has only two hidden layer with one different structure:

In the former, the amount of neurons in the first hidden layer is set to 2*(k+m), and (k+m) in the second hidden layer [28]. The initial values of the connection weights are randomly picked out in the range [−0.5, 0.5].

The networks will be trained by using the backpropagation algorithm with a sigmoidal activation function for both the hidden and output layers. The backpropagation algorithm is simple and easy to compute. It often converges rapidly to a local minimum, but it may not find a global minimum and in some cases, it may not converge at all [20]. In our experiments, the learning rate is set to 0.00001 and the goal error is set to .00001 in the first module whereas the number of training iterations is 1000 to the three modules, to the second module the learning rate is set to .001 and the goal error is set to .001, and to the third module the learning rate is set to .0001 and the goal error is set to .0001. In this paper we are going to work with the ensemble neural networks that have different parameters from learning rate and goal error, with the purpose of that are expert modules within the operation of the ensemble neural network.

3.2 Database Fingerprint

The Database used in our experiments was taken from The Biometric System Laboratory is active at the University of Bologna since 1993 and is supported by DEIS (Department of Electronics, Computer sciences and Systems) and

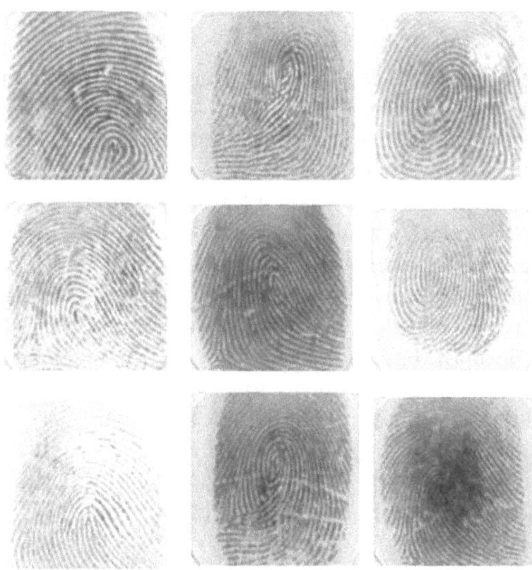

Fig. 2. Sample images from FCV 2000 database; Images; all the samples are from different fingers and are ordered by persons.

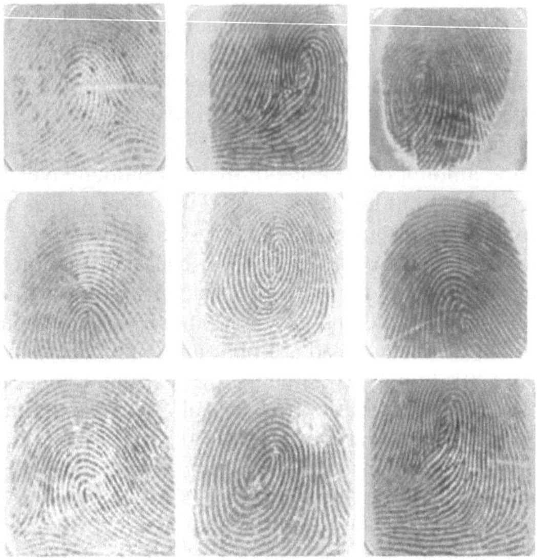

Fig. 3. Sample images from FCV 2000 database; all the samples are from different fingers and are ordered by persons.

Computer Science degree course - Cesena. The main research efforts are devoted to fingerprint and face recognition and to performance evaluation of biometric systems. Collaborations with industrial partners ensure that the research activities are linked to real applications [8].

Database features:

- The fingerprints were acquired by using a low-cost optical sensor
- The image size is 300 x 300.
- The fingerprints are mainly from 20 to 30 year-old students (about 50% male).
- Up to four fingers were collected for each volunteer (forefinger and middle finger of both the hands).
- The images were taken from untrained people in two different sessions and no efforts were made to assure a minimum acquisition quality.
- All the images from the same individual were acquired by interleaving the acquisition of the different fingers (e.g., first sample of left forefinger, first sample of right forefinger, first sample of left middle, first sample of right middle, second sample of the left forefinger,).
- The presence of the fingerprint cores and deltas is not guaranteed since no attention was paid on checking the correct finger centering.
- The sensor platens were not systematically cleaned (as usually suggested by the sensor vendors).
- The acquired fingerprints were manually analyzed to assure that the maximum rotation is approximately in the range [-15°, 15°] and that each pair of impressions of the same finger have a non-null overlapping area.

Sample images from the FCV 2000 database are shown in figures 2 and 3; are from different fingers and are ordered by persons [9].

4 Proposed Ensemble Neural Network for Fingerprint Recognition

The Proposed Ensemble Network Neural for Fingerprint Recognition in this paper consists of three main modules, in which each module in turn consists of a set of neural networks trained with the same data (fingerprints), which provides the ensemble neural network proposed architecture that is shown in Figure 4. The integration of responses is performed with a type-1 and type-2 fuzzy logic that is optimized using a GA [15].

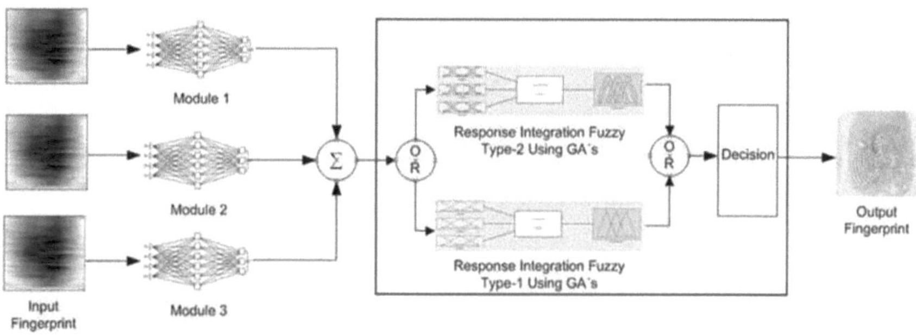

Fig. 4. Ensemble Neural Network Proposed for Fingerprint Recognition using GA's for Optimize Type-1 and Type-2 Fuzzy Rules.

4.1 Chromosome of the Genetic Algorithms for Response Integration Fuzzy Logic Type-1

The GA deals with coding the problem using chromosomes to optimize the fuzzy rules of the type-1 and type-2 fuzzy logic system used as response integration methods in ensemble neural networks [16], [17]. In each generation, the GAs builds a new population using genetic operators such as crossover and mutation, that is, members with higher fitness values are more likely to survive and participate in mating crossover operations. After a number of generations, the populations contains members with better fitness values in which this is analogous to Darwinian models of evolution by random mutation and natural selection. GAs is sometimes referred to as methods of population based optimization that improves performance by upgrading entire populations rather than individual members.

In this paper, the GAs are used to optimize the fuzzy rules of the type-1 and type-2 fuzzy logic systems as response integration methods in ensemble neural networks [9]. The GA deals with coding the problem using chromosomes to optimize the fuzzy rules of the type-1 and type-2 fuzzy logic system [11]. The

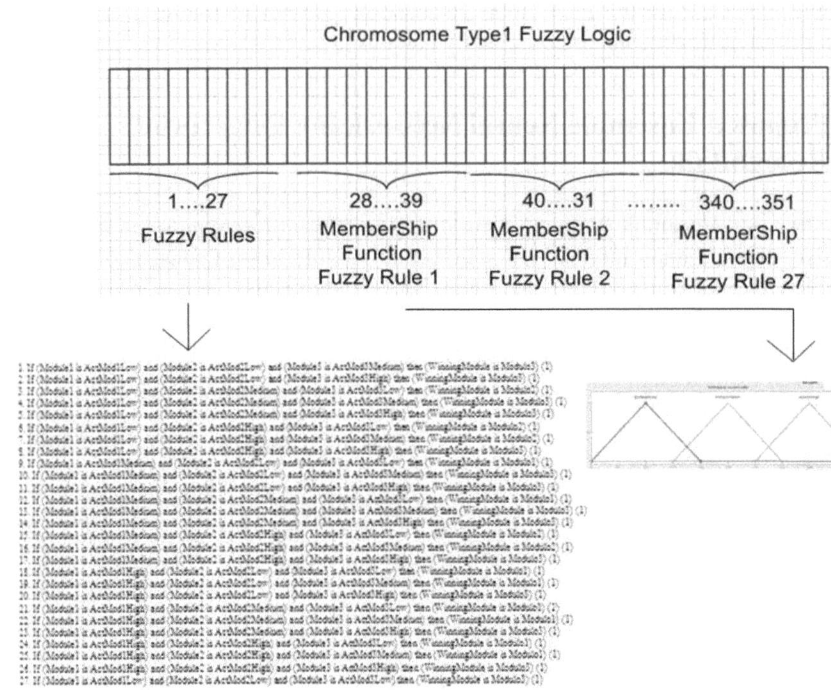

Fig. 5. The Chromosome Structure of Type-1 Fuzzy Logic to optimize the fuzzy rules.

Chromosome Structure used to optimize the fuzzy rules of the type-1 fuzzy system for response integration method in ensemble neural networks is shown in the figure 5. The first 27 bits represent the Fuzzy Rules of the fuzzy inference system. The fuzzy inference system has three input linguistic variables, and three memberships function each and one output with three membership function. The output membership functions are defined as Activation Low, Activation Medium and Activation High. In the same way the bits from 28 to the 39 represent the first fuzzy rule activation for three input membership function and one output membership function, the bits from the 40 to the 31 represent the second fuzzy rule activation, and the last bits from the 340 to the 351 represent the last fuzzy rule activation (27).

4.2 Chromosome of the Genetic Algorithms for Response Integration Fuzzy Logic Type-2

The GAs are used to optimize the fuzzy rules of the type-2 fuzzy System as response integration method in ensemble neural networks [28]. The GA deals with coding the problem using chromosomes to optimize the fuzzy rules of the type-2 fuzzy logic system. The Chromosome Structure used to optimize the fuzzy rules of the type-2 fuzzy system for response integration method in ensemble neural networks is shown in the figure 6.

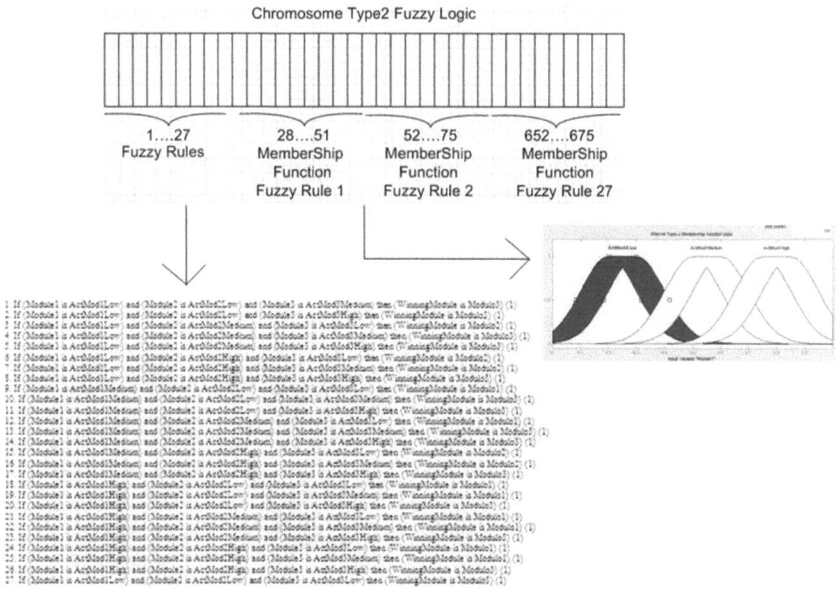

Fig. 6. The Chromosome Structure of Type-1 Fuzzy Logic to optimize the fuzzy rules.

The first 27 bits represent the Fuzzy Rules of the fuzzy inference system. The fuzzy inference system has three input linguistic variables, and sixth memberships function each and one output with six membership functions. The output membership functions are defined as Activation Low, Activation Medium and Activation High. The bits from 28 to the 51 represent the first fuzzy rule activation for three input membership function and one output membership function the membership functions designed using the editor of IT2FUZZY fuzzy logic toolbox, which was developed by our group [1], [2], the bits from the 52 to the 75 represent the second fuzzy rule activation, and the bits from the 625 to the 675 represent the last fuzzy rule activation (27).

4.3 Simulation Results Response Integration Type-1 and Type-2 Fuzzy Logic

Once the Ensemble Neural Network is trained, the fuzzy inference system integrates the outputs of the modules. We used the same 80 persons images to which we had applied different levels of noise with blur motion, both the type-1 and type-2 fuzzy inference system give an answer for the stage of the final decision, and show the result if the fingerprint input was recognized. We show in Figures 7 and 8 the simulation results using the response integration type-1 and type-2 fuzzy inference system obtained for the genetic algorithm. In the left side of the figure is the fingerprint input person with different level noise 10, 30, 50 and 90 Distance Pixels, and the right side the response integration shown the fingerprint output person.

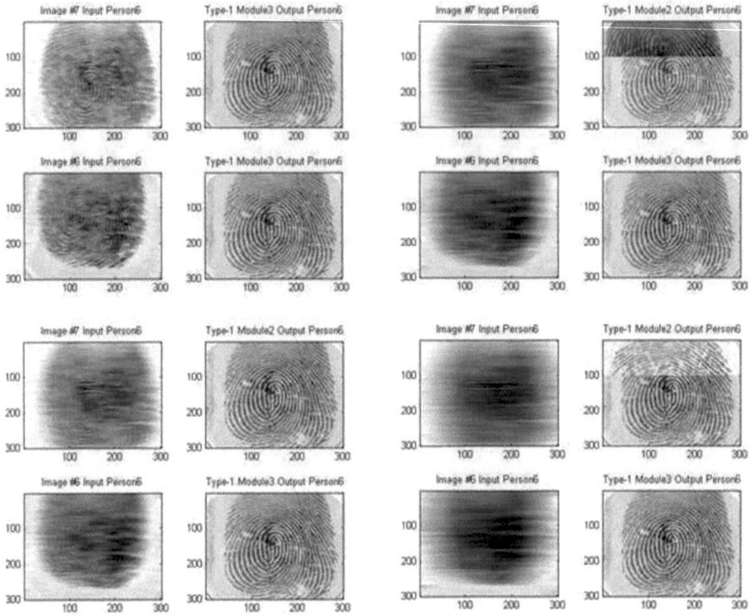

Fig. 7. Simulation Results Blur Motion Type-1 Fuzzy Logic Level Noise 10, 30 50, and 90 Distance.Pixels.

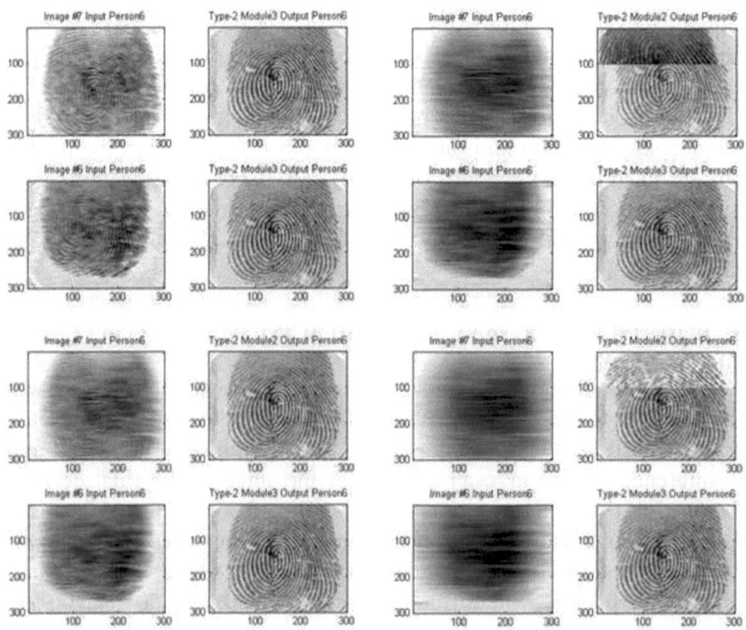

Fig. 8. Simulation Results Type-2 Fuzzy Logic Blur Motion Level Noise 10, 30, 50 and 90 Distance Pixels..

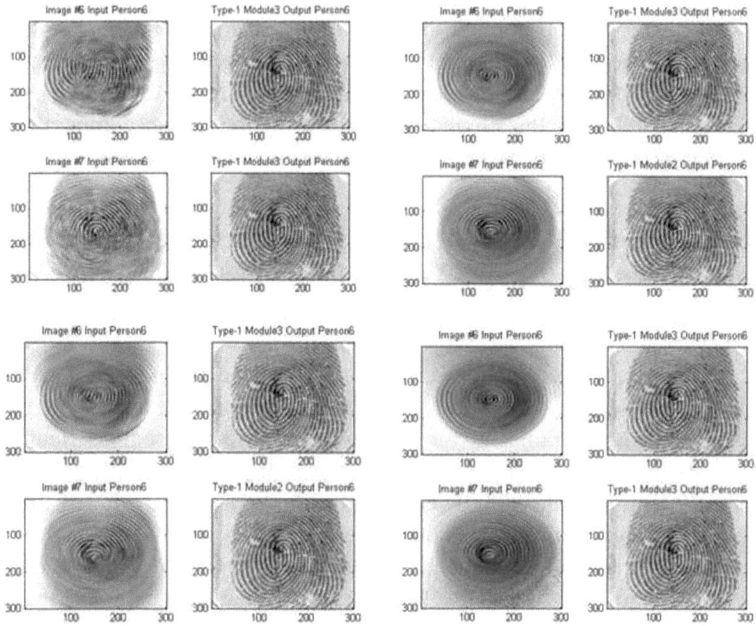

Fig. 9. Simulation Results Type-1 Fuzzy Logic Blur Radial Level Noise 10, 30, 50 and 90 Distance Pixels.

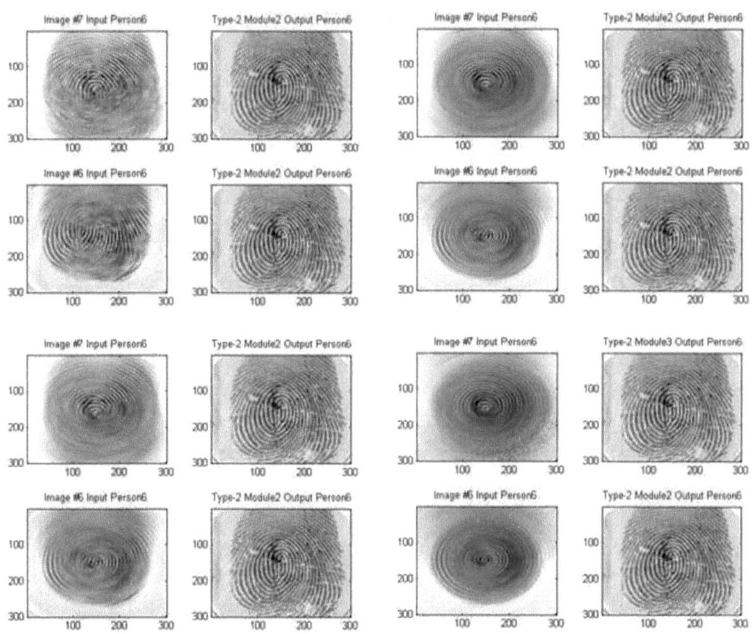

Fig. 10. Simulation Results Type-2 Fuzzy Logic Blur Radial Level Noise 10, 30, 50 and 90 Distance Pixels.

The Same way we had applied different levels of noise with blur radial, both the type-1 and type-2 fuzzy inference system gives an answer for the stage of the final decision, and show the result if the fingerprint input was recognized. We show in Figures 9 and 10 the simulation results using the response integration with type-1 and type-2 fuzzy inference systems obtained using the genetic algorithm.

5 Comparison of Results between Response Integration Fuzzy Logic Type-1 and Type-2 of the Ensemble Neural Network

The Experimental Results we had between Response Integration Type-1 and Type-2 Fuzzy System were obtained first using the same parameters of each module of the ensemble neural network, for example learning rate, and goal error, and after using different parameters for each module of the Ensemble Neural Networks, for module 1 the used parameters learning rate =.001, Goal Error=.001, for the module 2 learning rate =.0001, Goal Error=.0001, for the module 3 the learning rate =.0001, and Goal Error=.00001.

It increases the level noise to identification, the level noise blur motion and blurs radial increasing to 90 distances pixels. The experimental results obtained with blur motion level noise and blur radial level noises are shown in tables 1 and 2.

Table 1. Comparison Type-1, Type-1 Modified Parameters each module, Type-2 and Type-2 Modified Parameters each module of the Ensemble Neural Networks using blur motion level noise.

Response Integration Method	Recognition Rate	Identification Rate with Blur Motion Noise Distance Pixels					
		10	20	30	40	50	90
Type-2 using GA's with Mod. Par. Each Module	80/80 100%	100%	98.75	97.5	91.25	88.75	68.75
Type-2 using GA's	80/80 100%	97.5	90	87.5	83.75	78.75	55
Type-1 using GA's with Mod. Par. Each Module	80/80 100%	100	98.75	96.25	90	88.75	68.75
Type-1 using GA's	80/80 100%	97.5	93.75	90	85	77.5	53.75

Table 2. Comparison between Type-1, Type-1 Modified Parameters each of the module, Type-2 and Type-2 Modified Parameters each of the Ensemble Neural Networks using blur radial level noise.

Response Integration Method	Recognition Rate	Identification Rate with Blur Motion Noise Distance Pixels 10	20	30	40	50	90
Type-2 using GA's with Mod. Par. Each Module	80/80 100%	100	96.25	93.75	88.75	85	67.5
Type-2 using GA's	80/80 100%	98.75	93.75	90	81.25	71.25	50
Type-1 using GA's with Mod. Par. Each Module	80/80 100%	100	96.25	95	91.25	82.5	66.25
Type-1 using GA's	80/80 100%	98.75	93.75	90	82.5	71.25	48.75

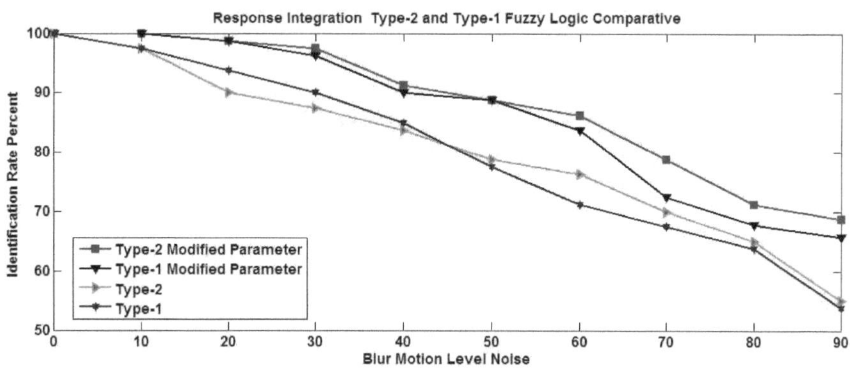

Fig. 11. Comparison with different Level Noise Blur Motion between Type-2 and Type-1 Modified Parameter Ensemble Neural Network, and Type-2 and Type-1 without Modified Parameter Fuzzy Inference System Using Gas to optimized the fuzzy rules.

In figure 11 we show Comparison between Type-1, Type-1 Modified Parameters each module, Type-2 and Type-2 Modified each of the module as Response Integration method in Ensemble Neural Networks, using GAs with blur motion level noise.

In figure 12 we show a comparison between Type-1, Type-1 with the modified Parameters in each module, Type-2 and Type-2 Modified each of the module as Response Integration method in Ensemble Neural Networks, using GAs with blur radial level noise.

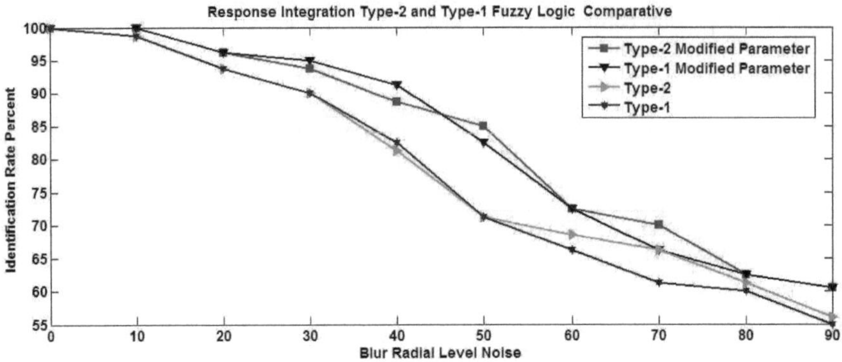

Fig. 12. Comparison with different level noise Blur Radial between Type-2 and Type-1 Modified Parameter Ensemble Neural Network, and Type-2 and Type-1 without Modified Parameter Fuzzy Inference System Using Gas to optimized the fuzzy rules.

6 Conclusions

Based on the experimental results, we can conclude that when the parameters of each module are changed of the Ensemble Neural Network significantly improved results which had obtained when they used the same parameters for each module.

Using a genetic algorithm to optimize the membership function of the type-1 and type-2 fuzzy system, as response integration of the output ensemble neural networks for the fingerprints, we obtain good results, mainly for the blur radial and blur motion level noise.

The difference between the type-1 fuzzy and type-2 fuzzy inference systems for response integration of ensemble neural networks is appreciated when we use blur motion level noise. We can conclude that the behavior can improve; we think that there is an advantage in using a type-2 fuzzy inference system to manage the uncertainty of the knowledge base in pattern recognition problems.

When we use a genetic algorithm to optimize the fuzzy rules of the type-1 and type-2 fuzzy logic, as response integration of the output ensemble neural networks for the fingerprints, the best results was obtained with blur motion level noise.

Future work will include, using different parameters in each module of the ensemble neural networks, other training methods for the neural networks, using wavelets like feature extraction, and other methods of the image compression, with the goal of improving the identification rate.

Acknowledgments

We would like to express our gratitude to the CONACYT, Universidad Autonoma de Baja California and Tijuana Institute of Technology for the facilities and resources granted for development of this research project.

References

[1] Castro, J.R., Castillo, O., Melin, P., Martinez, L.G., Escobar, S., Camacho, I.: Building Fuzzy Inference Systems with Interval Type-2 Fuzzy Logic Toolbox Number 1 in Studies in Fuzziness and Soft Computing, 1st edn., vol. 6, pp. 53–62. Springer, Germany (2007)

[2] Castro, J.R., Castillo, O., Melin, P.: An Interval Type-2 Fuzzy Logic Toolbox for Control Applications. In: Proc. FUZZ-IEEE 2007 (2007)

[3] Chang, S., Greenberg, S.: Fuzzy Measures and Integrals: Theory and Applications, pp. 415–434. Physica-Verlag, NY (2003)

[4] Cunningham, P.: Overfitting and Diversity in Classification Ensembles based on Feature Selection, TCD Computer Science Technical Report, TCD-CS-2000-07

[5] Grabisch, M., Murofushi, T., Sugeno, M.: Fuzzy Measures and Integrals: Theory and Applications, pp. 348–373. Physica-Verlag, NY (1989)

[6] Grabisch, M.: A new algorithm for identifying fuzzy measures and its application to pattern recognition. In: Proc. of 4th IEEE Int. Conf. on Fuzzy Systems, Yokohama, Japan, pp. 145–150 (1995)

[7] Gutta, S., Huang, J., Takacs, B., Wechsler, H.: Face Recognition Using Ensembles of Netrworks. In: 13th International Conference on Pattern Recognition (ICPR 1996), Vienna, Austria, vol. 4, p. 50 (1996)

[8] Maio, D., Maltoni, D., Cappelli, R., Wayman, J.L., Jain, A.K.: FVC2004: Third Fingerprint Verification Competition. In: Zhang, D., Jain, A.K. (eds.) ICBA 2004. LNCS, vol. 3072, pp. 1–7. Springer, Heidelberg (2004)

[9] Maltoni, D., Maio, D., Jain, A.K., Prabhakar, S.: The full FVC2000 and FVC2002 databases are available in the DVD included. In: Handbook of Fingerprint Recognition. Springer, New York (2003)

[10] MATLAB Trade Marks, ©1994-2007 by the MathWorks, Inc.

[11] Mostafa Abd Allah, M.: Artificial Neural Networks Fingerprints Authentication with Clusters Algorithm. Informatica 29, 303–307 (2005)

[12] Melin, P., Castillo, O.: Hybrid Intelligent Systems for Pattern Recognition Using Soft Computing. Springer, Heidelberg (2005)

[13] Melin, P., Castillo, O.: Hybrid Intelligent Systems for Pattern Recognition using Soft Computing. Springer, Heidelberg (2005), ISBN 3-540-24121-3

[14] Melín, P., González, F., Martínez, G.: Pattern Recognition Using Modular Neural Networks and Genetic Algorithms. In: ICAI 2004, pp. 77–83 (2004)

[15] Melín, P., Mancilla, A., Lopez, M., Solano, D., Soto, M., Castillo, O.: Pattern Recognition for Industrial Security using the Fuzzy Sugeno Integral and Modular Neural Networks. In: WSC11 11th Online World Conference on Soft Computing in Industrial Applications, September 18-October 6 (2006)

[16] Melín, P., Urias, J., Solano, D., Soto, M., Lopez, M., Castillo, O.: Voice Recognition with Neural Networks, Type-2 Fuzzy Logic and Genetic Algorithms. Engineering Letters 13(2), 108–116 (2006)

[17] Mendoza, O., Melin, P., Licea, G.: Modular Neural Networks and Type-2 Fuzzy Logic for Face Recognition. In: Reformat, M. (ed.) Proceedings of NAFIPS 2007, San Diego, June 2007, vol. 1 (2007), pages CD Rom

[18] Nemmour, H., Chibani, Y.: Neural Network Combination by Fuzzy Integral for Robust Change Detection in Remotely Sensed Imagery. EURASIP Journal on Applied Signal Processing 14, 2187–2195 (2005)

[19] Opitz, D.W.: Feature Selection for Ensembles. In: Sixteenth National Conference on Artificial/ Intelligence (AAAI), Orlando, FL, pp. 379–384 (1999)

[20] Opitz, D., Maclin, R.: Popular Ensemble Methods: An Empirical Study. Journal of Artificial Intelligence Research 11, 169–198 (1999)

[21] Opitz, D.W., Shavlik, J.W.: Generating accurate and diverse members of a neural network ensemble. In: Touretzky, D.S., Mozer, M., Hasselmo, M. (eds.) Advances in Neural Information Processing Systems 8, pp. 535–541. MIT Press, Cambridge (1996)

[22] Sharkey, A.C.: Modularity, combining and artificial neural nets, Connection Science, vol. 8, pp. 299–313 (1996)

[23] Sharkey, A.C.: On combining artificial neural nets, Connection Science, vol. 8, pp. 299–313 (1996)

[24] Urias, J., Solano, D., Soto, M., Lopez, M., Melín, P.: Type-2 Fuzzy Logic as a Method of Response Integration in Modular Neural Networks. In: ICAI 2006, pp. 584–590 (2006)

[25] Zadeh, L.A.: Fuzzy Logic. Computer 1(4), 83–93 (1998)

[26] Oh, J.-H., Kang, K.: Experts or an Ensemble A Statistical Mechanics Perspective of Multiple Neural Network Approaches, Korea, 790–784.

[27] Jain, A., Bolle, R., Pankanti, S.: Biometrics - Personal Identification in Networked Society. Kluwer Academic Publisher, Dordrecht (1999)

[28] López, M., Melin, P., Castillo, O.: Optimization of Response Integration with Fuzzy Logic in Ensemble Neural Networks Using Genetic Algorithms. Soft Computing for Hybrid Intelligent Systems, 129–150 (2008)

[29] Fingerprint Identification Systems, FBI, http://www.fbi.gov/hq/cjisd/ident.pdf

[30] Liu, Y., Yao, X., Higuchi, T.: Designing Neural Network Ensembles by Minimizing Mutual Information. In: Proceedings of the IASTED International on Conference Modeling, Simulation and Optimization MSO 2003, Banff, Canada, July 2-4, pp. 167–172 (2003)

Part III
Neural Network Models and Time Series Prediction

Ensemble Neural Networks with Fuzzy Integration for Complex Time Series Prediction

Martha Pulido, Alejandra Mancilla, and Patricia Melin

Tijuana Institute of Technology, Tijuana, B. C., México
epmelin@hafsamx.org

Abstract. In this paper we describe the application of the architecture for an ensemble neural network for Complex Time Series Prediction. The time series we are considering is the Mackey-Glass, and we show the results of some simulations with the ensemble neural network, and its integration with the methods of average, weighted average and Fuzzy Integration. Simulation results show very good prediction of the ensemble neural network with fuzzy integration.

1 Introduction

Time series predictions are very important because we can analyze past events to know the possible behavior of futures events and thus can take preventive or corrective decisions to help avoid unwanted circumstances.

The choice and implementation of an appropriate method of prediction has always been a major issue for enterprises that seek to ensure the profitability and survival of business. The predictions give the company the ability to make decisions in the medium and long term, and due to the accuracy or inaccuracy of data could mean predicted growth or profits and financial losses.

It is very important for companies to know the behavior that will be the future development of their business, and thus be able to make decisions that improve the company's activities, and avoid unwanted situations which in some cases can lead to the company's failure.

2 Time Series and Prediction

A time-series is defined as a sequence of observations on a set of values that takes a variable (quantitative) at different points in time. The time series are widely used today because organizations need to know the future behavior of certain phenomena in order to plan, prevent, and so on, their actions. That is, to predict what will happen with a variable in the future from the behavior of that variable in the past [1]. The data can behave in different ways over time, this may be a trend,

P. Melin et al. (Eds.): Bio-inspired Hybrid Intelligence Systems, SCI 256, pp. 143–155.
springerlink.com

which is the component that represents a long-term growth or decline in number over a period of time high. You can also have a cycle, which refers to the wave motion that occurs around the trend, or may not have a defined or random manner; there are seasonal variations (annual, biannual, etc.). , which is a behavior pattern that is repeated year after year at a particular time. [2].

The word "prediction" comes from the Latin prognosticum, which means I know in advance. Prediction is to issue a statement about what is likely to happen in the future, based on analysis and considerations of experiments. Making a forecast is to obtain knowledge about uncertain events that are important in decision-making [3]. Time series prediction tries to predict the future based on past data, it take a series of real data $x_t - n, \ldots x_t - 2, x_t - 1, x_t$ and then obtain the prediction of data $x_t + 1, x_t + 2 \ldots x_t + n$. The goal of time series prediction or a model is to observe the series of real data, so that future data may be accurately predicted. [4]

3 Neural Networks

Neural networks are composed of many elements (Artificial Neurons), grouped into layers and are highly interconnected (with the synapses), this structure has several inputs and outputs, which are trained to react (or give values) in a way you want to input stimuli (R values). These systems emulate in some way, the human brain. Neural networks are required to learn to behave (Learning) and someone should be responsible for the teaching or training (Training), based on prior knowledge of the environment problem [5].

Artificial neural networks are inspired by the architecture of the biological nervous system, which consists of a large number of relatively simple neurons that work in parallel to facilitate rapid decision-making [6].

A neural network (NN) is a system of parallel processors connected as a directed graph. Schematically each processing element (neuron) of the network is represented as a node. These connections establish a hierarchical structure that is trying to emulate the physiology of the brain as it looks for new ways of processing to solve real world problems. What is important in developing the techniques of NNs is that it is useful to learn behavior, recognize and apply relationships between objects and plots of real-world objects themselves. In this sense, artificial neural networks have been applied to many problems of considerable complexity. Its most important advantage is in solving problems that are too complex for conventional technologies, problems that have no solution or that the algorithm of the solution is very difficult to find [5].

4 Methods of Integration

There exists a diversity of methods of integration or aggregation of information, and we mention some of these methods below:

Integration by average: this method is used in the ensembles of networks. This integration method is the simplest and most straightforward, consists in the sum of the results generated by each module is divided by the sum of the number of modules, and the disadvantage is that there are cases in which the prognosis is not good.

Integration of Weighted Average: this method is an extension of the integration by average, with the main difference that the weighted average assigns importance weights to each of the modules. These weights are assigned to a particular module based on several factors; the most important is the knowledge product of experience. This integration method belongs to the well known aggregation operators.

Fuzzy logic was proposed for the first time in the mid-sixties at the University of California Berkeley by the brilliant engineer Lotfi A. Zadeh. Who proposed what it's called the principle of incompatibility: "As the complexity of system increases, our ability to be precise instructions and build on their behavior decreases to the threshold beyond which the accuracy and meaning are mutually exclusive characteristics." Then introduced the concept of a fuzzy set (Fuzzy Set), under which lies the idea that the elements on which to build human thinking are not numbers but linguistic labels. Fuzzy logic can represent the common knowledge that kind of language is mostly qualitative and not necessarily quantity in a mathematical language by means of fuzzy set theory and function characteristics associated with them. [7].

Fuzzy Set: Let X be a space of objects and x be generic element of X. A classical set$A$, $A \subseteq X$, is defined as a collection of elements or objects $x \in X$, such that each x can either belong or not belong to the set A. By defining a characteristic function for each element x in X, we can represent a classical set A by a set of ordered pairs $(x, 0)$ or $(x, 1)$, which indicates $x \notin A$ or $x \in A$, respectively. Unlike the aforementioned conventional set, a fuzzy set expresses the degree to which an element belongs to a set. Hence the characteristic function of a fuzzy set is allowed to have values between 0 and 1, which denotes the degree of membership of an element in a given set. The basic structure of a fuzzy inference system consists of three conceptual components: a rule base, which contains a selection of fuzzy rules; (or dictionary), which defines the membership functions used in the fuzzy rules; and a reasoning mechanism, which performs the inference produce (usually the fuzzy reasoning). [8].

5 Genetic Algorithms

Genetic algorithms were introduced by the first time by a professor of the University of Michigan named John Holland [9]. A genetic algorithm, it is a mathematical highly parallel algorithm that transforms a set of mathematical individual objects with regard to the time using operations based on evolution. The Darwinian laws of reproduction and survival of the fittest can be used, and after having appeared of natural form a series of genetic operations of between(among) that stands out the sexual recombination [10,11]. Each of these mathematical objects is in the habit of being a chain of characters (letters or numbers) of fixed length that adjusts to the model of the chains of chromosomes, and one associates to them with a certain mathematical function that reflects the fitness.

6 Problem Statement and Proposed Method

This paper is concerned with the study of a fuzzy integration method that can be applied to ensemble neural networks with applications to complex time series, in addition to developing alternative methods for the integration of ensemble network as are the average and weighted average. Figure 1 presents the general architecture used in this work.

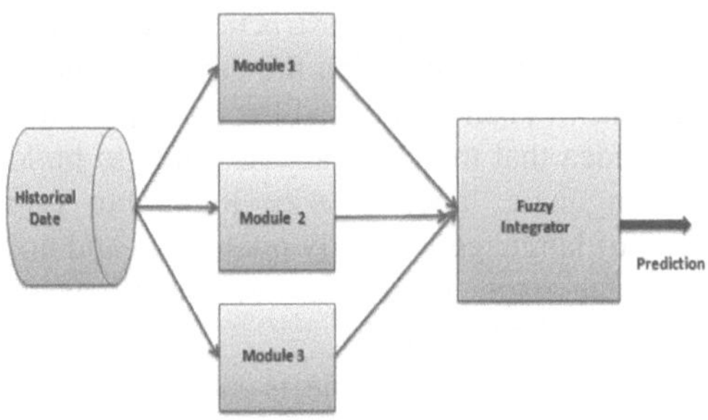

Fig. 1. General Architecture of the ensemble neural network.

Historical data of the Mackey-Glass time series was used for the ensemble neural network trainings, where each module was fed with the same information, to find a suitable architecture for a module of the ensemble will be same or very similar, unlike the modular networks, where each module is fed with different data, which leads to architectures that are not uniform. Integration by the average method was very easy to implement, just joining the results of each module and the result was divided by the number of elements, the main problem of this method is that if one of the modules produces an unfortunate result, it can greatly affect the result the integration. Integration by weighted average includes assigning values from 0 to 1, where the module that has the best prediction is the one that will have a greater weight. The network consists of three modules, the allocation of weights we used is 0.50 for the module that produces better results, 0.30 for second best and 0.20 for the best module, we do this only if the three modules meet the above conditions. Fuzzy integration for the Mackey-Glass time series was implemented in a system of traditional Mamdani fuzzy inference, which consists of three input variables (the results of each module of our ensemble network) and one output variable (the result of integration), is shown in Figure 2.

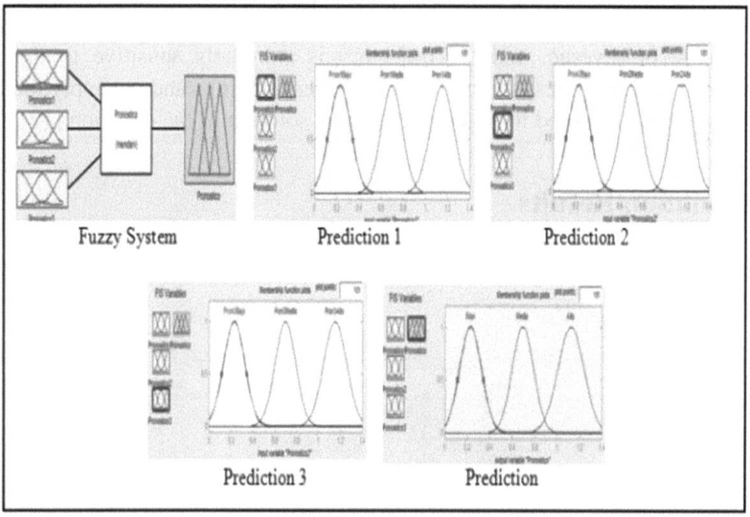

Fig. 2. Fuzzy Inference System.

Data of the Mackey-Glass time series was generated using equation (1). We are using 800 points. We use 70% of the data for the ensemble neural network trainings and 30% to test the network.

The Mackey-Glass Equation is defined as follows:

$$\dot{x}(t)=\frac{0.2x(t-\tau)}{1+x^{10}(t-\tau)}-0.1x(t) \tag{1}$$

Where it is assumed x (0) = 1.2, la t = 17, and x (t) = 0 for t <0. Figure 3 shows a plot of the time series for these parameter values.

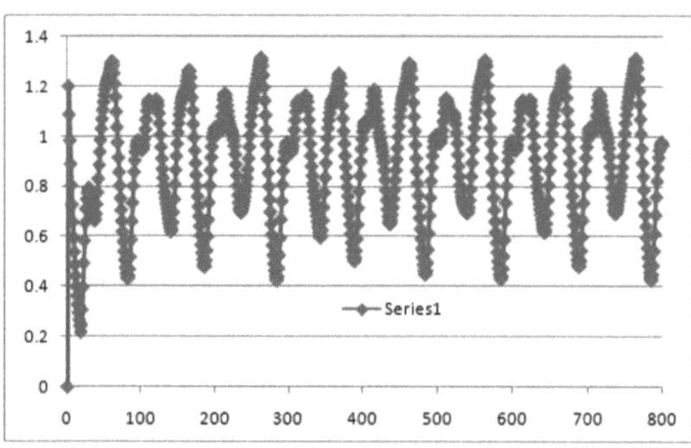

Fig. 3. Series data Mackey-Glass.

This time series is chaotic, and there is no clearly defined period. The series does not converge or diverge, and the trajectory is extremely sensitive to the initial conditions. The time series is measured in number of points, and we apply the fourth order Runge-Kutta method to find the numerical solution of the equation [12].

7 Simulation Results

The architecture of the ensemble network that produced the best results is the one shown in Figure 4.

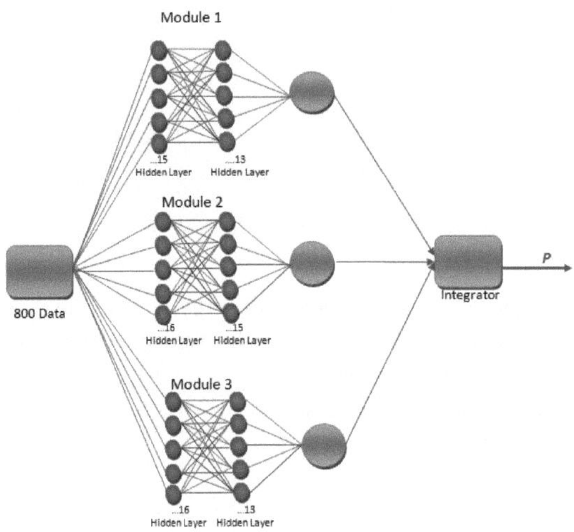

Fig. 4. Network Architecture.

Table 1. Results of Training.

	Number of Layers	Number of Neurons	Number of Delays	**Target Error**	Number Max Epoch	Epoch	Time	**Error by Module**	Error Total
Training1	1	37 35 35	3	0.00000001 0.00001 0.00001	16000 20000 20000	4 1 1	1 Seg 1 Seg 1 Seg	0.0144 0.0617 0.0417	0.2703
Training2	1	37 35 35	3	0.00000001 0.00001 0.00001	16000 20000 20000	4 2 1	1 Seg 1 Seg 1 Seg	0.0983 0.074 0.024	0.1803
Training3	1	29 30 31	3	0.000001 0.00001 0.00001	6000 2000 5000	3 1 1	1 Seg 1 Seg 1 Seg	0.0777 0.0347 0.0517	0.1296
Training4	1	25 28 29	3	0.0000001 0.0000001 0.00000001	6000 3000 5000	2 2 4	1 Seg 1 Seg 1 Seg	0.1072 0.0356 0.0973	0.1752
Training5	1	27 30 28	3	0.0000001 0.0000001 0.00000001	6000 3000 5000	2 2 7	1 Seg 1 Seg 1 Seg	0.0894 0.0577 0.058	0.1664

Table 2. Results of Training.

	Number of Layers	Number of Neurons	Number of Delays	Target Error	Number Max Epoch	Epoch	Time	Error by Module	Total Error
Training1	2	29,10 18,18 22,12	3	0.00000001 0.00000001 0.000001	11000 20000 10000	7 8 5	1 Seg 1 Seg 1 Seg	0.076 0.0425 0.1157	0.1571
Training2	2	29,12 18,16 20,17	3	0.00000001 0.00000001 0.000001	11000 20000 10000	7 9 8	1 Seg 1 Seg 1 Seg	0.0533 0.0479 0.0795	0.1277
Training3	2	18,12 18,17 21,19	3	0.0000001 0.000000001 0.000001	7000 1000 9000	9 11 7	1 Seg 1 Seg 1 Seg	0.0924 0.0543 0.1672	0.2025
Training4	2	18,13 15,18 15,18	3	0.0000001 0.000000001 0.000000001	7000 1000 1000	10 26 9	1 Seg 1 Seg 1 Seg	0.0579 0.0798 0.073	0.162
Training5	2	15,13 16,15 11,13	3	0.00000001 0.00000001 0.0000001	6000 3000 10000	11 8 10	1 Seg 1 Seg 1 Seg	0.0196 0.0116 0.0098	0.0136

In this architecture we used two layers in each module. In module 1, in the first layer we used 15 neurons and 13 neurons in second layer. In module 2 we used 15 neurons in the first layer and 13 neurons in the second, and in module 3 we used 16 neurons in the first layer and 13 neurons in the second. The training method used was the Levenberg-Marquardt (LM); we also applied 3 delays to the network.

The best training was the one shown in row number 3, using 1 layer in each of the modules of the ensemble network, implementing 3 delays in the network, the training method used was Levenberg-Marquardt (LM), (as shown in Table 1) where the total error of this training was: 0.1296.

The best training was the one shown in row number 5 using 2 layers in each of the modules of the ensemble network, implementing 3 delays in the network, the training method used was Levenberg-Marquardt (LM), (as shown in Table 2) where the total error of this training was: 0.01396.

Table 3. Results of Training.

	Number of Layers	Number of Neurons	Number of Delays	Target Error	Number Max Epoch	Epoch	Time	Error by Module	Total Error
Training1	1 y 2	10,17 30 18,11	3	0.0000001 0.0000001 0.00000001	6000 3000 5000	6 1 10	1 Seg 1 Seg 1 Seg	0.0622 0.0295 0.0255	0.1002
Training2	1 y 2	13,12 30 10,17	3	0.00000001 0.00000001 0.00000001	6000 2000 5000	8 11 9	1 Seg 1 Seg 1 Seg	0.027 0.1237 0.0613	0.1712
Training3	1 y2	13,15 31 15,10	3	0.00000001 0.00000001 0.0000001	6000 2000 3000	11 4 11	1 Seg 1 Seg 1 Seg	0.1134 0.1016 0.0407	0.2285
Training4	1 y 2	31 31 15,10	3	0.00000001 0.00000001 0.00000001	2000 2000 5000	55 4 9	1 Seg 1 Seg 1 Seg	0.0848 0.1058 0.0926	0.2215
Training5	1 y 2	23 25 10,14	3	0.000000001 0.0000001 0.0000001	6000 1000 3000	44 1 8	1 Seg 1 Seg 1 Seg	0.0144 0.0617 0.417	0.0899

Table 4. Results of the Integration Methods

	Average Integration	Weighted Average Integration	Fuzzy Integration
Trainingl	0.0417	0.0558	0.1151
Training2	0.317	0.0436	0.0715
Training3	0.025	0.0461	0.0789
Training4	0.0205	0.0245	0.0694
Training5	0.0261	0.0326	0.0787

Table 5. Results of the Integration Methods

	Average Integration	Weighted Average Integration	Fuzzy Integration
Training1	0.0396	0.0737	0.0813
Training2	0.0388	0.0439	0.0815
Training3	0.0583	0.075	0.0826
Training4	0.0272	0.0355	0.0899
Training5	0.0106	0.0126	0.0708

The best training was the one shown in row number 5 varying between 1 and 2 layers in each of the modules of the ensemble network, implementing 3 delays in the network, the training method used was Levenberg-Marquardt (LM), (as shown in Table 3) where the total error of this training was: 0.01396.

The best result is training number 4 (as shown in Table 4), where the integration method used was average integration; and we obtained an error of 0.0205, with weighted average integration we obtained an error of 0.0264 and with fuzzy integration we obtained an error of 0.0694.

The best method of integration for this architecture was the average integration with an error of 0.0205.

The best result is training 5 (as shown in Table 5), where the integration method used was average integration; we obtained an error of 0.0106, with weighted average integration we obtained an error of 0.0126 and with fuzzy integration we obtained an error of 0.0708.

Table 6. Results of the Integration Methods

	Average Integration	Weighted Average Integration	Fuzzy Integration
Training1	0.0264	0.037	0.0681
Training2	0.0258	0.053	0.0728
Training3	0.03	0.0307	0.0738
Training4	0.0312	0.0297	0.0963
Training5	0.0241	0.0348	0.0751

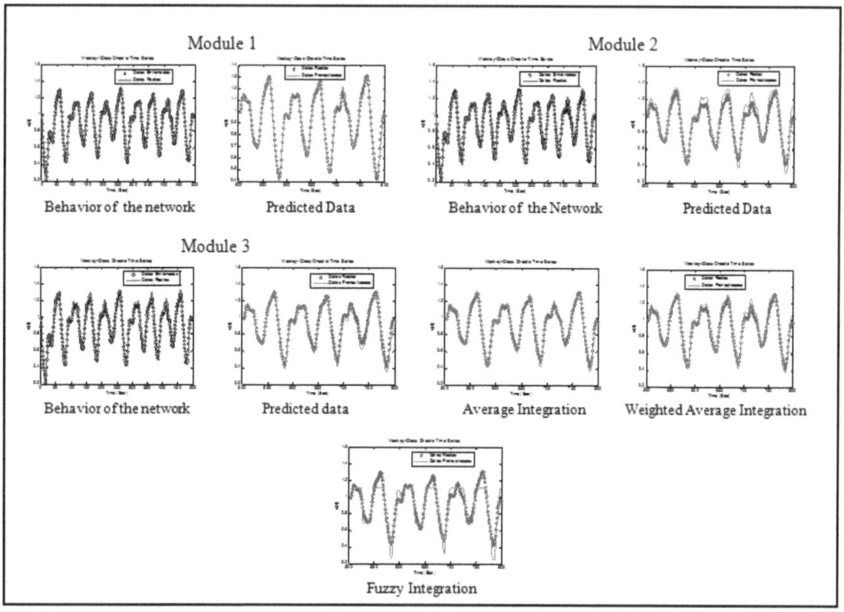

Fig. 5. Simulation results of training 5

The best result is training 5 with average integration method we obtained an error of 0.0241, for the weighted average integration was training4 the best, we obtained an error of 0.0297, and with fuzzy integration training1 was the best, we obtained an error of 0.0681. (as shown in Table 6).

Where the best method for this architecture was the average integration with an error of 0.0241.

The best result obtained for the Mackey-Glass time series was number 5 using 2 layers in each of the modules of the ensemble network, implementing 3 delays, the training method used was Levenberg-Marquardt (LM), (as shown in Table 2 and in Figure1).

To improve Fuzzy Integration of the Mackey-Glass time series we implemented a genetic algorithm to optimize the membership functions of the fuzzy system. Where the objective function is defined to minimize the error predition:

$$f_1 = \left(\sum_{i=1}^{297} |a_i - x_i| \right) / 297$$

$$f_2 = \left(\sum_{i=1}^{297} |b_i - x_i| \right) / 297 \qquad (2)$$

$$f_3 = \left(\sum_{i=1}^{297} |c_i - x_i| \right) / 297$$

$$f = (f_1 + f_2 + f_3)/3$$

Where $a, b, and\ c,$ correspond to prediction1, prediction2 and prediction3, respectively; these are used as inputs for the fuzzy integration system, X represents real data, f_1, f_2 and f_3 calculates the error of module1, module2, module3 respectively and f is the total prediction error.

The corresponding chromosome structure is shown in Figure 6.

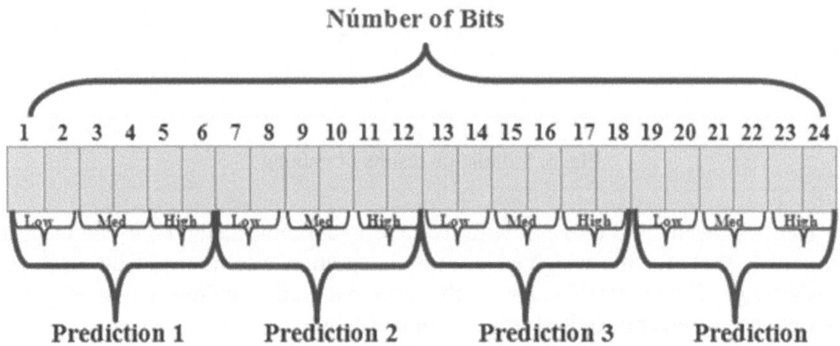

Fig. 6. Chromosome Structure

Table 7. Genetic algoritm results.

Num.	Ind.	Gen	Selection	GGP	Mutation	Pm	Crossover	Pc	Duration	Error
1	100	100	rws	0.85	mutbga	0.06	xovsp	0.09	02:24:42	0.024193
2	100	100	rws	0.85	mutbga	0.05	xovsp	0.09	02:12:50	0.023475
3	100	100	rws	0.85	mutbga	0.05	xovsp	0.08	02:12:53	0.022164
4	100	100	rws	0.85	mutbga	0.05	xovsp	0.09	02:16:09	0.023475
5	100	100	rws	0.85	mutbga	0.06	xovsp	1	02:51:08	0.022998
6	100	100	rws	0.85	mutbga	0.08	xovsp	1	04:23:09	0.022384
7	100	100	rws	0.85	mutbga	0.07	xovsp	0.7	03:19:50	0.024335
8	100	100	rws	0.85	mutbga	0.06	xovsp	0.8	03:44:56	0.02134
9	100	100	rws	0.85	mutbga	0.06	xovsp	0.7	03:44:06	0.022868
10	70	100	rws	0.85	mutbga	0.5	xovsp	1	02:33:57	0.020462
11	40	100	rws	0.85	mutbga	0.08	xovsp	1	02:31:59	0.022594
12	100	80	rws	0.85	mutbga	0.6	xovsp	1	02:00:12	0.021316
13	100	100	rws	0.85	mutbga	0.06	xovsp	0.5	02:15:20	0.027117
14	100	100	rws	0.85	mutbga	0.8	xovsp	0.6	01:55:07	0.024051
15	100	100	rws	0.85	mutbga	0.07	xovsp	0.5	02:04:21	0.020308
16	100	100	rws	0.85	mutbga	0.5	xovsp	0.06	01:57:08	0.023819
17	20	100	rws	0.85	mutbga	0.6	xovsp	1	00:23:13	0.024894
18	40	100	rws	0.85	mutbga	0.8	xovsp	1	00:46:37	0.023954
19	50	70	rws	0.85	mutbga	0.5	xovsp	0.7	00:41:29	0.022985
20	30	80	rws	0.85	mutbga	0.5	xovsp	0.7	00:28:15	0.026477

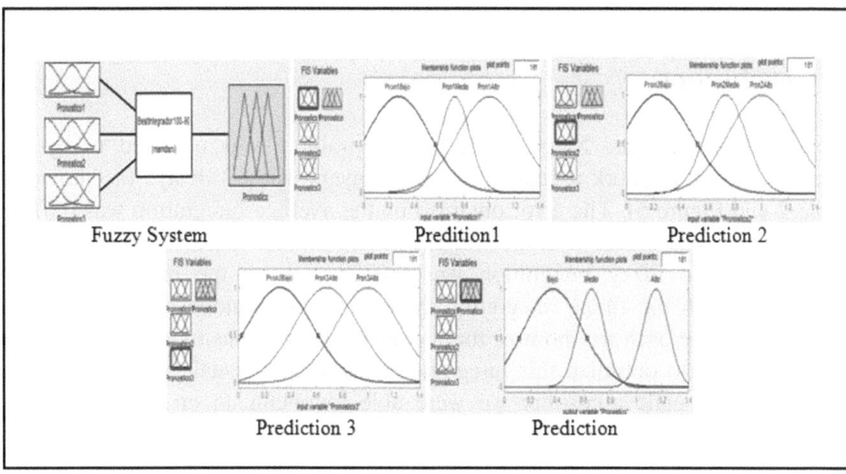

Fig. 7. Fuzzy system.

Table 6 shows the genetic algorithm results for training 5 (as shown in Table 2 and in Figure 1). Where the prediction error is of 0.020308.

The fuzzy integration system generated by the genetic algorithm is represented in figure 7.

Comparing Figure 2, which belongs to the base fuzzy integrator (not optimized) with Figure 7, which belongs to the fuzzy integrator generated by the genetic algorithm, it can be seen that the parameters of membership functions are different and this helps to improve the error prediction.

In Figure 8, we illustrate a graphical representation of the fuzzy integration obtained by the fuzzy integrator generated by the genetic algorithm.

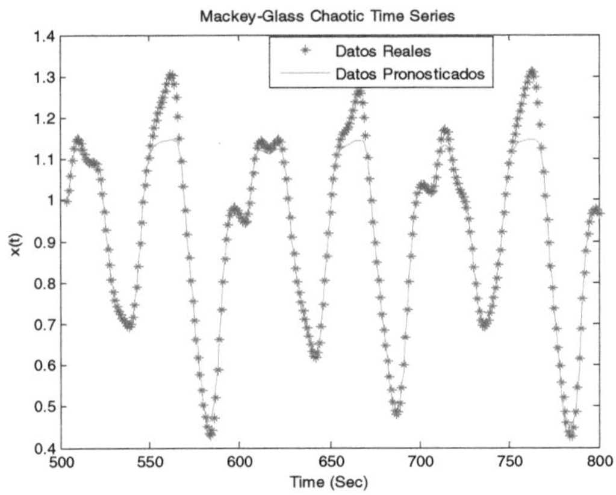

Fig. 8. Prediction with the optimized fuzzy integration

8 Conclusions

The best result for the Mackey-Glass series training was obtained using an ensemble neural network architecture with 2 layers using 3 delays (as shown in Table 2 and Figure 5). The error obtained by the average integration was 0.0106, the error obtained by the average weighted integration was 0.0126 and the error obtained by the fuzzy integration was 0.0708. Since the results were not satisfactory with the fuzzy integration system, it was decided to implement an evolutionary approach to optimize the membership functions of this system. The method chosen to optimize this integration system is a genetic algorithm. After applying the genetic algorithm we were able to obtain an error of 0.020308, improving the results obtained with the non optimized system.

Acknowledgment

We would like to express our gratitude to the CONACYT, Tijuana Institute of Technology for the facilities and resources granted for the development of this research.

References

[1] Vásquez, G., Muñoz, T.J.: Time Series. Catholic University of the Holy Concepción, Chile (2001), http://zip.rincondelvago.com/?00033622 (December 05, 2008)

[2] Arellano, M.: Introduction to time series analysis (2001), http://ciberconta.unizar.es/LECCION/seriest/100.HTM (September 06, 2008)

[3] Davey, N., Hunt, S., Frank, R.: Time Series Prediction and Neural Networks. University of Hertfordshire, Hatfield (1999) (December 04, 2008)

[4] Plummer, E.A.: Time series forecasting with feed-forward neural Networks: uidelines and limitations. University of Wyoming (July 2000) http://www.karlbranting.net/papers/plummer/Paper_7_12_00.htm (January 20, 2009)

[5] Neural Network Tutorial (December 05, 2008), http://www.answermath.com/neural-networks/tutorial-esp-2-concepto.htm

[6] Multaba, I.M., Hussain, M.A.: Application of Neural Networks and Other Learning. Technologies in Process Engineering, Process Engineering. Imperial Collage Press (December 03, 2001)

[7] Basic Concepts of Fuzzy Logic (September 10, 2008), http://www.tdx.cbuc.es/TESIS_UPC/AVAILABLE/TDX0207105105056//04Rpp04de11.pdf

[8] Jang, J.S.R., Sun, C.T., Mizutani, E.: Neuro-Fuzzy and Soft Computing. Prentice Hall, Englewood Cliffs (1996)

[9] Holland, J.: Adaptation in natural and artificial systems. University of Michigan Press (January 05,1975)

[10] Golberg, D.: Genetic Algorithms in search, optimization and machine learning. Addison Wesley, Reading (1989) (January 20, 2009)

[11] Yen, J., Langari, R.: Fuzzy Logic: intelligence, control and Information. Prentice Hall, Englewood Cliffs (1999) (January 17, 2009)

[12] MATLAB, Historical Mackey Glass data (2007) (September 1, 2008)

Prediction of the MXNUSD Exchange Rate Using Hybrid IT2 FLS Forecaster

Gerardo M. Mendez and Angeles Hernandez

Instituto Tecnologico de Nuevo Leon, Av. Eloy Cavazos #2001,
CP 67170, Cd. Guadalupe, NL, México
gmm_paper@yahoo.com.mx, ahernandez@yturria.com.mx

Abstract. This paper presents a novel application of the interval type-1 non-singleton type-2 fuzzy logic system (FLS) for one step ahead prediction of the daily exchange rate between Mexican Peso and US Dollar (MXNUSD) using the recursive least-squared (RLS)-back-propagation (BP) hybrid learning method. Experiments show that the exchange rate is predictable. A non-singleton type-1 FLS and an interval type-1 non-singleton type-2 FLS, both using only BP learning method, are used as a benchmarking systems to compare the results of the hybrid interval type-1 non-singleton type-2 FLS (RLS-BP) forecaster.

Keywords: Type-2 fuzzy inference systems, type-2 neuro-fuzzy systems, hybrid learning, uncertain rule-based fuzzy logic systems.

1 Introduction

Interval type-2 fuzzy logic systems (IT2 FLS) constitute an emerging technology [1]. As in hot strip mill process [2], [3], in autonomous mobile robots [4], and in plant monitoring and diagnostics [5], [6], financial systems are characterized by high uncertainty, nonlinearity and time varying behavior [7], [8]. This makes it very difficult to forecast financial variables such as exchange rate, closing prices of stock indexes and inflation. The ability of comprehending as well as predicting the movements of economic indices could result in significant profit, stable economic environments and careful financial planning. Neural networks are very popular in financial applications and a lot of work has been done in exchange rate and stock markets predictions described elsewhere [9]–[11].

T2 fuzzy sets (FS) let us model the effects of uncertainties, and minimize them by optimizing the parameters of the IT2 FS during a learning process [12]–[14]. Although some econometricians [7], [8], claim that the raw data used to train and test the FLS forecasters should not be directly used in the modelling process, since it is time varying and contains non-stationary noise, they were used by us to directly train the IT2 FLS. The inputs were modeled as type-1 (T1) non-singletons.

P. Melin et al. (Eds.): Bio-inspired Hybrid Intelligence Systems, SCI 256, pp. 157–164.
springerlink.com

The interval type-1 non-singleton type-2 fuzzy logic system (IT2 NSFLS-1) forecaster accounts for all of the uncertainties that are present in the initial antecedents and consequents of the fuzzy rule base.

2 MXNUSD Exchange Rate Prediction

The data used to train the three forecasters cover a period of nine years and four and a half months from 01/02/1997 to 06/17/2006 whereas the test data cover five months from 06/18/2006 to 10/12/206. The daily closing price of MXNUSD exchange rate was found on the Web site: http://pacific.commerce.ubc.ca/xr/. It was a set of $N=$ 2452 data, $s(1)$, $s(2)$, …, $s(2452)$. It is assumed that the noise-free sampled MXNUSD exchange rate, $s(k)$, can be corrupted by uniformly distributed stationary additive noise $n(k)$, so that

$$x(k)=s(k)+n(k) \qquad k=1,2,\ldots,2452 \ . \tag{1}$$

and that signal to noise ratio (SNR) is equal to 0 dB. In this application $n(k)$ was fixed to 0. Figure 1, shows the trend of the raw noise-free data $s(k)$. The first 2352 noisy-free data were used for training, and the remaining 100 data were used for testing the forecasters. Four antecedents were used as noise-free inputs, $x(k-3)$, $x(k-2)$, $x(k-1)$ and $x(k)$, to predict the noise-free output, $y=x(k+1)$.

2.1 IT2 NSFLS-1 Design

The architecture of the IT2 NSFLS-1 was established in such a way that its parameters were continuously optimized. The number of rule-antecedents (inputs) was fixed to four. Each antecedent-input space was divided into three FS, resulting in 81 rules. Gaussian primary membership functions (MF) of uncertain means were chosen for the antecedents. Each rule of the IT2 NSFLS-1 was characterized by 12 antecedent MF parameters (two for left-hand and right-hand bounds of the mean and one for standard deviation, for the FS of each of the four antecedents), and two consequent parameters (one for left-hand and one for right-hand end points of the consequent type-1 FS), giving a total of 14 parameters per rule. Each input value had one standard deviation parameter, giving 4 additional parameters.

The resulting IT2 NSFLS-1 used type-1 (T1) non-singleton fuzzification, join under maximum t-conorm, meet under product t-norm, product implication, and center-of-sets type-reduction.

The IT2 NSFLS-1 was trained using two main learning mechanisms: the back-propagation (BP) method for tuning of both antecedent and consequent parameters, and the hybrid training method using recursive least-squared method (RLS) for tuning of consequent parameters as well as the BP method for tuning of antecedent parameters. In this work, the former is referred to as IT2 NSFLS-1 (BP), and the latter as hybrid IT2 NSFLS-1 (RLS-BP).

Fig. 1. The daily closing price of MXNUSD exchange rate.

2.2 IT2 Fuzzy Rule Base

The IT2 NSFLS-1 fuzzy rule base consists of a set of IF-THEN rules that represents the model of the system. This IT2 NSFLS-1 has four inputs $x_1 \in X_1$, $x_2 \in X_2$, $x_3 \in X_3$, $x_4 \in X_4$ and one output $y \in Y$. The rule base has M = 81 rules of the form

$$R^i : IF \quad x_1 is \tilde{F}_1^i \quad and \quad x_2 is \tilde{F}_2^i, \quad and \quad x_3 is \tilde{F}_3^i \quad and \quad x_4 is \tilde{F}_4^i, \; THEN \quad y \quad is \quad \tilde{G}^i . \tag{2}$$

where i = 1,2,3,...,81 rules; $\tilde{F}_1^i, \tilde{F}_2^i, \tilde{F}_3^i$ and $\tilde{F}_4^i$ are the antecedent type-2 fuzzy sets, and $\tilde{G}^i$ is the consequent type-2 fuzzy set of rule i.

2.3 Antecedent Membership Functions

The primary MF for each antecedent is an IT2 FS described by Gaussian primary MF with uncertain means:

$$\mu_k^i(x_k) = \exp\left[-\frac{1}{2}\left[\frac{x_k - m_k^i}{\sigma_k^i}\right]^2 \right]. \tag{3}$$

where $m_k^i \in \left[m_{k1}^i, m_{k2}^i\right]$ is the uncertain mean, and σ_k^i is the standard deviation, with k =1,2,3,4 (the number of antecedents) and i = 1,2,...,81 (the number of M rules). Table I shows the values of the three FSs for each antecedent.

Table 1. Intervals of uncertainty for all inputs.

	m_{k1}	m_{k2}	σ_k^i
1	7.58	7.78	1.04
2	9.68	9.88	1.04
3	11.68	11.88	1.04

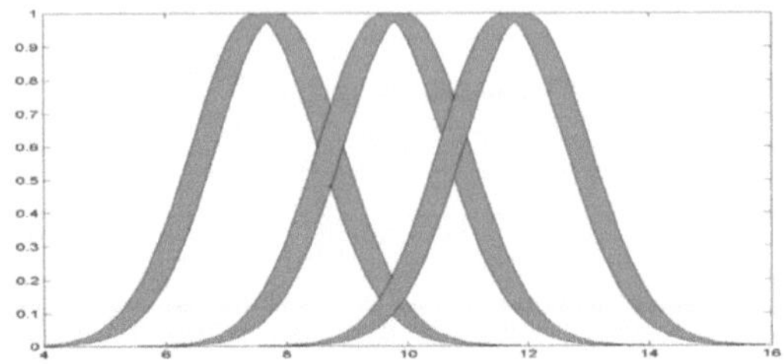

Fig. 2. Thee FSs for each of the four antecedents.

Fig. 2 shows the initial FSs of each antecedent, being the same for all inputs.

2.4 Consequent Membership Functions

The primary membership function for each consequent is a Gaussian with uncertain means, as defined in equation (3). Because the centre-of-sets type-reducer replaces each consequent set $\tilde{G}^i$ by its centroid, then y_l^i and y_r^i (the M left-points and right points of interval consequent centroids) are the consequent parameters.

The initial values of the centroid parameters y_l^i and y_r^i are such that the corresponding MFs uniformly cover the output space, from 4.0 to 16.0.

3 Modeling Results

One T1 NSFLS and two IT2 NSFLS-1 FLSs were trained and used to predict the daily MXNUSD. The BP training method was used to train both the base line T1 NSFLS and IT2 NSFLS-1 forecasters, while the hybrid RLS-BP method was used to train the hybrid IT2 NSFLS-1 forecaster. For each of the two training methods, BP and RLS-BP, we ran 25 epoch computations; 1138 parameters were tuned using 2352 input-output training data pairs per epoch.

The performance evaluation for the learning methods was based on root mean-squared error (RMSE) benchmarking criteria as in [1]:

$$RMSE_{S,2}(*) = \sqrt{\frac{1}{n}\sum_{k=1}^{n}\left[Y(k) - f_{nS,2}\left(\mathbf{x}^{(k)}\right)\right]^2} \quad . \tag{4}$$

where $Y(k)$ is the output from the 100 testing data pairs, $f_{nS,2}\left(\mathbf{x}^{(k)}\right)$ is the predicted output of the type-1 non-singleton type-2 forecaster, evaluated using the input data $\mathbf{x}^{(k)}$, $k = 1,2,\ldots,100$, $RMSE_{S,2}(*)$ stands for $RMSE_{S,2}(BP)$ [the RMSE of the IT2 NSFLS-1 (BP)], and for $RMSE_{S,2}(RLS-BP)$ [the RMSE of the IT2 NSFLS-1 (RLS-BP)]. The same formula was used in order to calculate the $RMSE_{S,1}(BP)$ of a base line T1 NSFLS (BP).

Fig. 3, shows RMSEs of the two IT2 NSFLS-1 systems and the base line T1 NSFLS for 25 epochs of training. For the T1 NSFLS (BP), there is a substantial performance improvement at epoch two, and after twenty-five epochs of training, the performance is still improving. For the IT2 NSFLS-1 (BP), there is small improvement in performance after five epochs of training. For the case of the hybrid IT2 NSFLS-1 (RLS-BP), its performance still improves after twenty-five epoch of training.

Observe that the IT2 NSFLS-1 (BP) only has better performance than T1 NSFLS-1 during the first twenty epochs of training.

In other hand, after one epoch of training, the hybrid IT2 NSFLS-1 (RLS-BP) has better performance than both T1 NSFLS (BP) and IT2 NSFLS-1 (BP) for all the twenty-five epochs. The hybrid IT2 NSFLS-1 (RLS-BP) outperforms both the T1 NSFLS (BP) and the IT2 NSFLS-1 (BP) at each epoch. The average of the $RMSE_{S,2}(BP)$ is 92.86 percent of the average of the $RMSE_{S,1}(BP)$, representing an improvement of 7.14 percent. The $RMSE_{S,2}(RLS-BP)$ improves the $RMSE_{S,1}(BP)$ by 83.52 percent and the $RMSE_{S,2}(BP)$ by 82.25 percent, as can be observed in Table 2.

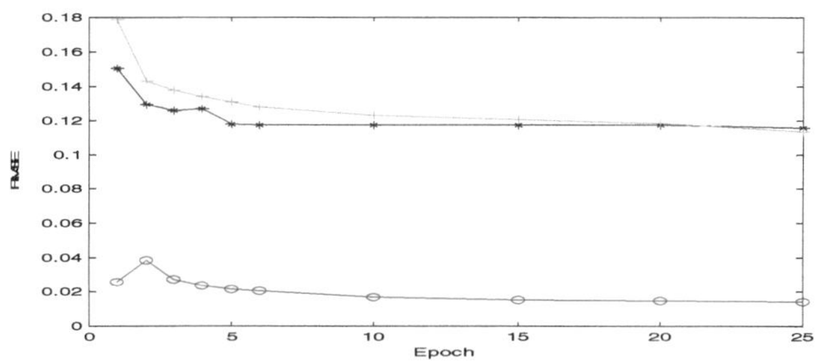

Fig. 3. RMSE errors of the three forecasters - TI NSFLS (BP) (+), IT2 NSFLS-1 (BP) (*), IT2 NSFLS-1 (RLS-BP) (o).

Table 2. Performance efficiency of the three fuzzy forecasters.

Forecasters	Performance ratio
T1 NSFLS (BP)	1.00
IT2 NSFLS (BP)	7.14
IT2 NSFLS-1 (RLS-BP)	82.25

A comparison between the predicted exchange rate after one epoch and twenty-five epochs of training using T1 NSFLS (BP), and the real price , are depicted in Figures 4 and 5, respectively. Figures 6 and 7 respectively show the IT2 NSFLS-1 (BP) predictions after one epoch and twenty-five epochs of training. Observe that IT2 NSFLS-1 (BP) prediction after epoch one takes better into account the variations of the real data than its counterpart T1 NSFLS-1 (B). The predicted

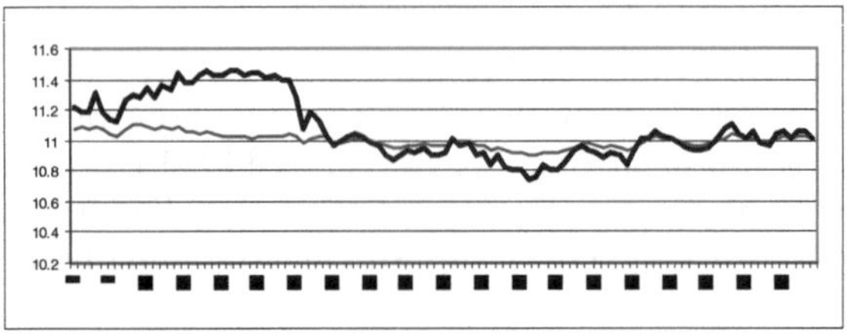

Fig. 4. (Blue) Real historical and (Red) predicted prices of MXNUSD exchange rate. Predictions of the T1 NSFLS (BP) forecaster after one epoch of training.

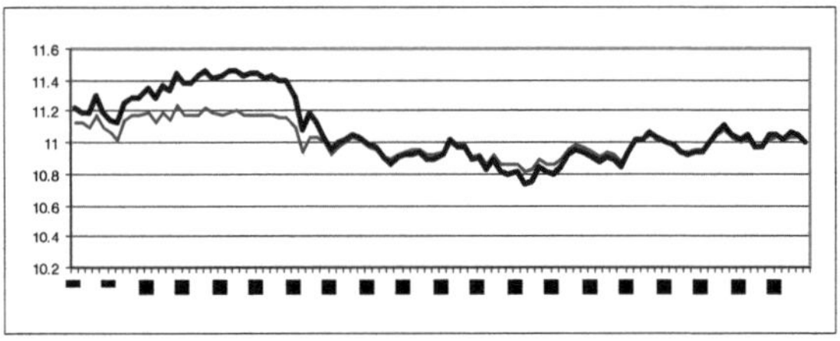

Fig. 5. (Blue) Real historical and (red) predicted prices of MXNUSD exchange rate. Predictions after twenty-five epochs of training of T1 NSFLS (BP).

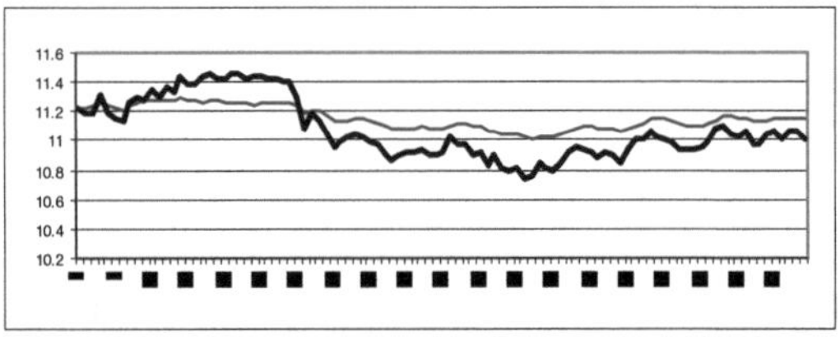

Fig. 6. (Blue) Real historical and (red) predicted prices of MXNUSD exchange rate. Predictions after one epoch of training of IT2 NSFLS-1 (BP).

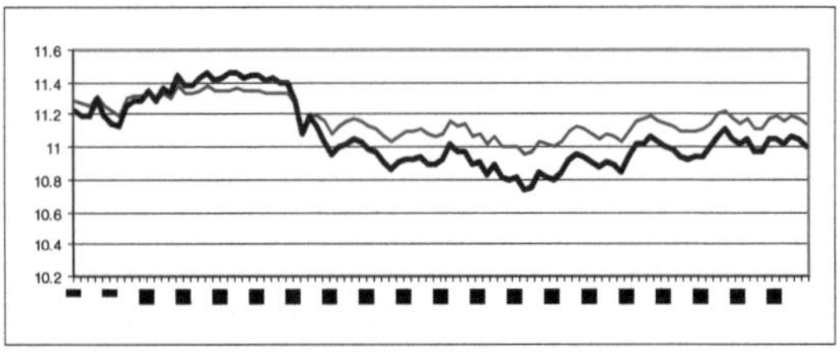

Fig. 7. (Blue) Real historical and (red) predicted prices of MXNUSD exchange rate. Predictions after twenty-five epochs of training of IT2 SFLS-1 (BP).

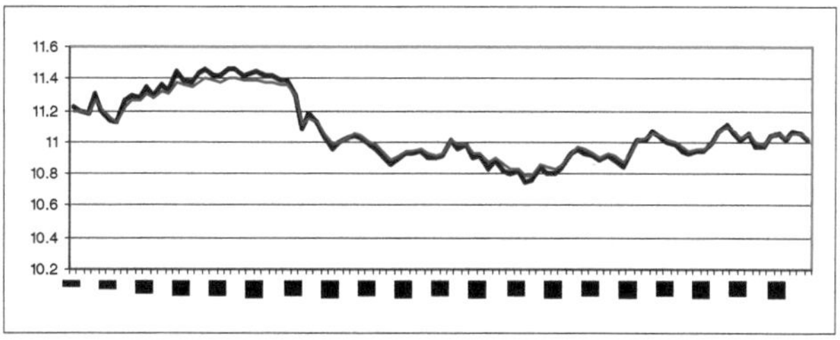

Fig. 8. (Blue) Historical and (red) predicted prices of MXNUSD exchange rate. Predictions after one epoch of training of hybrid IT2 NSFLS-1 (RLS-BP).

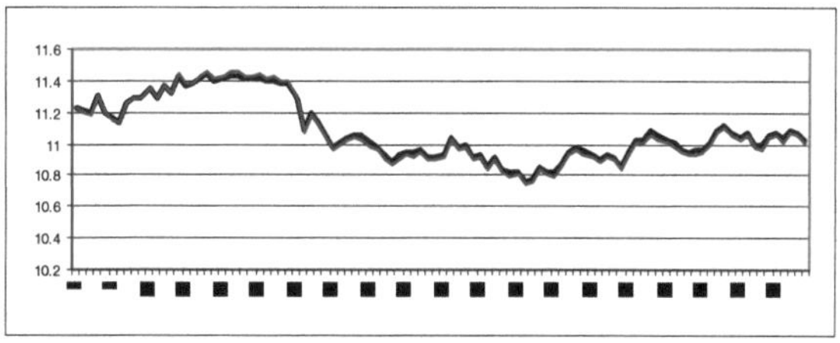

Fig. 9. (Blue) Historical and (red) predicted prices of MXNUSD exchange rate. Predictions after twenty-five epochs of training of hybrid IT2 NSFLS-1 (RLS-BP).

exchange rate using hybrid IT2 NSFLS-1 (RLS-BP) after one epoch and twenty-five epochs of training, are depicted in Figures 8 and 9, respectively. The predicted value is in all figures one step ahead in order to compare with its real value. After only one epoch of training, the hybrid IT2 NSFLS-1 (RLS-BP) predictions are very close to the real values, meaning that it is feasible its application to predict and to control critical systems where there is only one chance of training. The reason is that the uncertainties presented in the training data are almost totally incorporated into the 81 rules of the IT2 NSFLS-1 forecaster, and that the fast convergence of the predicted error is improved by the RLS method.

4 Conclusions

An IT2 NSFLS-1 using the hybrid learning method RLS-BP was tested and compared for forecasting the daily exchange rate between Mexican Peso and U.S. Dollar (MXNUSD. The results showed that the hybrid IT2 NSFLS-1 (RLS-BP) forecaster provided the best performance, and the base line T1 NSFLS-1 (BP) provided the worst performance. We conclude, therefore, that it is possible to directly use the daily data of exchange rate to train an IT2 NSFLS-1, in order to model and predict MXNUSD exchange rate one day in advance. It was observed that IT2 NSFLS-1 forecasters efficiently managed the raw historical data.

References

1. Mendel, J.M.: Uncertain rule-based fuzzy Logic systems: Introduction and New Directions. Prentice-Hall, Upper Saddle River (2001)
2. Mendez, G., Cavazos, A., Leduc, L., Soto, R.: Hot strip temperature prediction using hybrid learning for interval singleton type-2 FLS. In: Proceedings of the IASTED International Conference on Modelling and Simulation, pp. 380–385. Palm Springs, CA (2003)
3. Mendez, G.: Orthogonal-back propagation hybrid learning algorithm for type-2 fuzzy logic systems. In: IEEE Proceedings of the NAFIPS 2004 International Conference on Fuzzy Sets, Banff Alberta Canada, vol. 2, pp. 899–902 (2004)
4. Hagras, H.A.: A hierarchical type-2 fuzzy logic control architecture for autonomous mobile robots. IEEE Transactions on Fuzzy Systems 12(4), 524–539 (2004)
5. Castillo, O., Melin, P.: A new hybrid approach for plant monitoring and diagnostics using type-2 fuzzy logic and fractal theory. In: Proceedings of the International Conference FUZZ 2003, pp. 102–107 (2003)
6. Castillo, O., Huesca, G., Valdez, F.G.: Evolutionary computing for optimizing type-2 fuzzy systems in intelligent control of non-linear dynamic plants. In: IEEE Proceedings of the NAFIPS 2005 International Conference, pp. 247–251 (2005)
7. Anastasakis, L., Mort, N.: Prediction of the GSPUSD exchange rate using statistical and neural network models. In: Proceedings of the IASTED International Conference on Artificial Intelligence and Applications, Benalmadena, España, pp. 493–498 (2003)
8. Lendasse, A., de Boot, E., Wertz, V., Verleysen, M.: Non-linear financial time series forecasting – application to the bel 20 stock market index. European Journal of Economics and Social Systems 14, 81–91 (2000)
9. Ito, Y.: Neuro-fuzzy phillips-type stabilization policy for stock process and exchange-rates models. In: Proceedings of the IASTED International Conference on Artificial Intelligence and Applications, Benalmadena, España, pp. 499–504 (2003)
10. Saad, E.W., Prokhorov, D.V., Wunch II, D.C.: Comparative study of stock trend prediction using time delay, recurrent and probabilistic neural networks. IEE Trans. On Neural Networks 9, 1456–1470 (1998)
11. Leigh, W., Paz, M., Purvis, R.: An analysis of a hybrid neural network and pattern recognition technique for short term increases in the NYSE composite index. The International Journal of Management Science 30, 169–176 (2002)
12. Liang, Q., Mendel, J.M.: Interval type-2 fuzzy logic systems: theory and design. Trans. on Fuzzy Systems 8, 535–550 (2000)
13. John, R.I.: Embedded interval valued type-2 fuzzy sets. In: Proceedings of the 2002 IEEE International Conference on Fuzzy Systems, Honolulu, Hawaii, vol. 1&2, pp. 1316–1321 (2002)
14. Mendel, J.M., John, R.I.: Type-2 fuzzy sets made simple. IEEE Transactions on Fuzzy Systems 10, 117–127 (2002)

Discovering Universal Polynomial Cellular Neural Networks through Genetic Algorithms

Eduardo Gomez-Ramirez[1], Enrique Haro Sedeño[1], and Giovanni Egidio Pazienza[2]

[1] LIDETEA, POSGRADO E INVESTIGACION La Salle University, México, D.F.
egr@ci.ulsa.mx, enrique_haro77@yahoo.com.mx
[2] GRSI, Enginyeria i Arquitectura La Salle, Universitat Ramon Llull,
Quatre Camins 2, 08022 Barcelona, Spain
gpazienza@salle.url.edu

Abstract. There are different benchmarks to test the capabilities of artificial neural networks. The Game of Life is an algorithm that has a very interesting mathematical characterization because can realize universal computing in the sense of a Turing machine. In this paper a new model of Polynomial Cellular Neural Networks that simulates a semi-totalistic automata is presented with the learning design to compute the templates. In this case, the rules of the semi-totalistic automata used, "play" the Game of Life. With the simulations presented we show that the PCNN can realize universal computing.

Keywords: polynomial cellular neural networks, semi-totalistic automata, game of life.

1 Introduction

Since the original work of Leon Chua [1], the Cellular Neural Network was born as the union of the cellular automata and artificial neural network. Despite this origin, there are few papers that really combine both theories [2][3][4]. Some of them describe the relation in specific applications or dynamics such as the game of life [5]. The Game of Life is really a no-player game with very interesting and complex behavior that can be "played" with very simple rules [6]. However, it is important to note that this game only can be implemented using a multilayer CNN [7]. The single-layer space-invariant CNN model is not capable of classifying linearly non-separable data points. The response to this problem is the same as the historic discussion about the perceptrons many years ago published by Marvin Minsky [8]. A more recent description of this idea, with other approach, can be found in [9] using the concepts of linear threshold gates and how it is possible to solve problems with linearly nonseparable data points using quadratic threshold gates.

To improve the response of CNN some authors propose different options to the standard model, for example: removing the uniformity constraint on weight values [10], adding a trapezoidal activation function [11] or piecewise-linear discriminant

P. Melin et al. (Eds.): Bio-inspired Hybrid Intelligence Systems, SCI 256, pp. 165–175.
springerlink.com

functions [12]. A similar result can be obtained if nonlinear term is added. For example in [13] a new model of Polynomial Cellular Neural Network is proposed to solve the XOR (eXclusive OR) problem. The type and complexity of the term improves the mathematical representation capability of the network.

In this paper a new nonlinear term is added to the traditional discrete model with the result of a Polynomial Discrete time Cellular Neural Network (PDTCNN). This new model improves the capabilities of representation of the network with only one layer. To show the advantage of the algorithm a semitotalistic automaton, The Game of Life, was selected. This algorithm is traditional used in the area of cellular automata to show that some type of automata can realize universal computing.

The structure of the paper is the following: Section II explains the model of Polynomial Cellular Neural Network. Next section explains the learning methodology using Genetic Algorithm. Section IV shows some experiments and conclusions of the work

2 Polynomial Cellular Neural Networks

2.1 Cellular Neural Networks

Cellular Neural Network is a new paradigm [14] with local connections and geometrically placed, analog, 3-D regular processing arrays. In the most common model, cells are arranged in a square grid, and the state of each one depends directly on the input and the output of its 8 nearest cells. In this paper we deal with Discrete-Time CNNs (DTCNNs) [15], a CNN model in which only binary output values are allowed and the system is clocked. Its dynamic is described by the following equations

$$x_{ij}(n) = \sum_{C(k,l)\in N_r(i,j)} A(i,j;k,l)\, y_{kl}(n) + \sum_{C(k,l)\in N_r(i,j)} B(i,j;k,l)\, u_{kl} \tag{1}$$

$$y_{ij}(n+1) = f(x_{ij}(n)) = \begin{cases} 1 \ if \ x_{ij}(n) \geq 0 \\ -1 if \ x_{ij}(n) < 0 \end{cases} \tag{2}$$

where u_{ij}, x_{ij} and y_{ij} are: the input, the state and the output of the cell in position *(i,j)*, respectively; *Nr(i,j)* is the set of indexes corresponding to the cell *(i,j)* and its neighbors; finally, the matrices *A* and *B* contain the weights of the network and the value *z* is a bias. Usually, the weights of a CNN depend on the difference between cell indexes rather than on their absolute values, therefore the network is space-invariant. In this case the operation performed is defined by the set of weights *{A,B,z}* that is unique for the whole network and is called 'cloning template' (see [16]). Note that throughout this paper capital letters indicate matrices and lower case letters indicate scalar values. The Einstein notation - according to whom when an index variable appears twice in a single term, it implies that we are summing over all of its possible values - allows writing the Eq. (1) in a compact way as follows:

$$x^c(n) = a_d^c y^d(n) + b_d^c u^d(n) + z \tag{3}$$

2.2 Extension to the Polynomial Model

Despite their versatility, space-invariant Cellular Neural Networks are not able to solve non-linear separable problems, unless a multilayer structure is used [17]. This difficulty can be overcome by using a polynomial CNN (PCNN) [18, 19,20], whose discrete-time state equation is:

$$x^c(n) = a_d^c y^d(n) + b_d^c u^d(n) + z + g(u^d, y^d) \tag{4}$$

where a polynomial term $g(u^d, y^d)$ is added to the standard DTCNN model. The output is calculated as before through the Eq. (4). In [21] it is proved that a polynomial CNN allows to carry out the XOR operation - the most elemental non-linear separable task - whereas other authors have applied this model to real-life cases, like epilepsy seizure prediction [22]. The model proposed for $g(u^d, y^d)$ to simulate the semitotalistic automata is a second degree polynomial, whose general form is:

$$\begin{aligned} g(u^d, y^d) &= \sum_{m=0}^{2} (q_i)_d^c \left(u^d\right)^m \cdot (p_i)_d^c \left(y^d\right)^{2-m} \\ &= (q_0)_d^c \left(1^d\right) \cdot (p_0)_d^c \left(y^d\right)^2 + (q_1)_d^c \left(u^d\right) \cdot (p_1)_d^c \left(y^d\right) + (q_2)_d^c \left(u^c\right. \end{aligned} \tag{5}$$

where $(P_0,\ldots,P_2,Q_0,\ldots,Q_2)$ are matrices of dimensions $\in Nr(i,j)$.

This model was obtained using the same methodology proposed in [23]. In this paper, the optimal model of a polynomial artificial neural network to forecast nonlinear time series was computed using genetic algorithm.

3 Semitotalistic Cellular Automata: The Game of Life

In the GoL [24] the next state of the central cell depends only on its present state and the sum of its eight nearest neighbors. This kind of CA is called semitotalistic automata. If the state doesn't depend of the central cell is called totalistic automata. The GoL is a no-player game usually played on an infinite two-dimensional grid of square cells, and its evolution is determined only by the initial state.

Each cell interacts with its 8 neighbor cells, and at any fixed time it can be either alive (black cell) or dead (white cell). In each time step, the next state of each cell is defined by the following rules:

- Birth: a cell that is dead at time t becomes alive at time t + 1 only if exactly 3 of its eight neighbors were alive at time t;
- Survival: a cell that was living at time t will remain alive at t + 1 if and only if it had exactly 2 or 3 alive neighbors at time t.

The importance of the GoL comes from the fact that this structure has the same complexity as a universal Turing machine [25], therefore any conceivable algorithm can be computed by a GoL-based computer.

3.1 Representation of the Rules of Semitotalistic CA

The rules of any CA can be conveniently represented in a Cartesian system [26], as depicted in Fig. 1. Following the convention that a black pixel is +1 and a white pixel is -1, the x axis corresponds to the central pixel of the 3×3 Cellular Automaton, whereas the y axis is equal to the sum of the 8 outer neighbors. The rules of the CA are represented by associating to each point of the grid - corresponding to a pair (central pixel, sum of neighbors) - a symbol indicating the following state of the cell: a red cross represents a 1 (black) for the next state, a blue diamond represents a -1 (white). Thanks to this representation, it is evident that the GoL is not a linearly separable task. Moreover, these results can be summarized in table 1 and it is clear that the 18 combinations are necessary and sufficient to describe the GoL.

3.2 Remarks on the Relation between GoL and CNN

A fundamental aspect of CNN's is how to find a cloning template - that is, the set of weights - to perform a given operation. This issue, commonly called 'CNN learning', is crucial, and many efforts have been made to deal with it. It is possible to design explicitly a template only in simple cases (e.g. [27,28]) thanks to the formulation of local rules. However, often it is necessary to find the template through a supervised learning algorithm. What is more, gradient-based algorithms, successfully employed with a number of neural networks, do not work in general with CNNs [29]. Therefore, the only way to solve the problem is to apply global learning procedures, like genetic algorithms [30] or simulated annealing [31,32], in which the learning problem is reformulated as a general optimization problem. The cost function is usually the difference between the desired output and the one obtained using a certain configuration for the weights. Now, the point is to establish what parameters of the network must be learnt.

In our work we take into consideration a space-invariant linear DTCNN, whose model, described by the Eq. (3), comprises 19 free parameters (9 for the A matrix, 9 for the B matrix and the bias). When the second order polynomial of Eq. (5) is

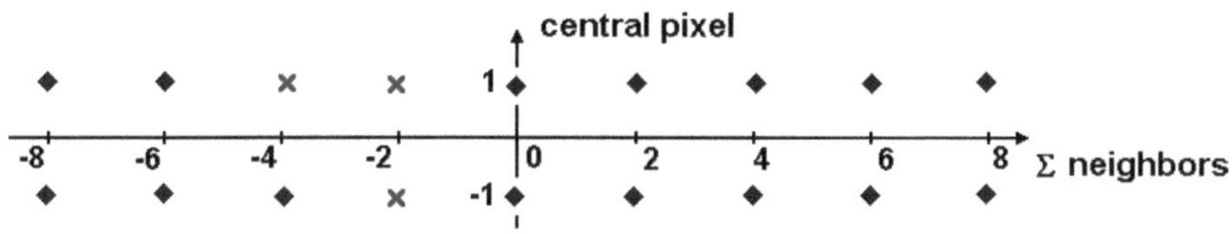

Fig. 1. The Game of Life: a red cross corresponds to a 1 (black cell) for the next state, and a blue diamond is a -1 (white cell).

Table 1. List of the 18 possible combinations for the GoL. (W-White, B-Black)

	Central Pixel	# Black neighbours	Output
1.	W	0	W
2.	W	1	W
3.	W	2	W
4.	W	3	B
5.	W	4	W
6.	W	5	W
7.	W	6	W
8.	W	7	W
9.	W	8	W
10.	B	0	W
11.	B	1	W
12.	B	2	B
13.	B	3	B
14.	B	4	W
15.	B	5	W
16.	B	6	W
17.	B	7	W
18.	B	8	W

added, there are 45 terms to consider (9 for each of the five matrices). This increment in parameters results in a huge search space, but making some preliminary consideration about the problem, the search space can be easily shrunk.

First, as detailed in previous section, the rules of the GoL are applied on a 3×3 cells square, and they focus only on the central cell and the number of black pixels in the neighborhood regardless their position. Consequently, a cloning template contains only matrices with central symmetry. Therefore, just two values of each matrix must be learnt: the central element (subindex *c*) and the peripheral element (subindex *p*). Second, we can state that the last term of Eq. (5) is constant because the input is binary; thus, it can be included in the bias. Also the value $\left(p_1\right)_d^c\left(1^d\right)$ belonging to the first term of Eq. (5) is constant, and then it can be ignored during the learning.

To sum up, the polynomial CNN model for the GoL can be written as:

$$x^c(n) = a_d^c y^d(n) + b_d^c u^d(n) + p_d^c y^d \cdot q_d^c u^d + r_d^c (y^d)^2 + z \qquad (6)$$

Where:

$$A = \begin{bmatrix} a_p & a_p & a_p \\ a_p & a_c & a_p \\ a_p & a_p & a_p \end{bmatrix} \quad B = \begin{bmatrix} b_p & b_p & b_p \\ b_p & b_c & b_p \\ b_p & b_p & b_p \end{bmatrix}$$

$$P = \begin{bmatrix} p_p & p_p & p_p \\ p_p & p_c & p_p \\ p_p & p_p & p_p \end{bmatrix} \quad Q = \begin{bmatrix} q_p & q_p & q_p \\ q_p & q_c & q_p \\ q_p & q_p & q_p \end{bmatrix} \quad R = \begin{bmatrix} r_p & r_p & r_p \\ r_p & r_c & r_p \\ r_p & r_p & r_p \end{bmatrix}$$

In conclusion, thanks to the remarks made in this section, we realized that only 11 parameters $\{a_c, a_p, b_c, b_p, p_c, p_p, q_c, q_p, r_c, r_p, i\}$ must be learnt, which is a great improvement with respect to the initial hypothesis of 46 free parameters. A similar scheme can be used to learn semitotalistic cellular automata with other complexity (more than 2).

4 Genetic Algorithm

GA is a search technique inspired on some of the evolution mechanisms that are observed in nature such as: inheritance, mutation, selection and crossover [33][34]. There are many versions of this algorithm since the first model was proposed by John Holland in 1975 [35].

In this work a modified version of the work presented in [36] is used to train the PDTCNN. In this version we take advantage of the parallel computing toolbox of Matlab to improve the response of the previous model.

4.1 Crossover

The crossover operator is characterized by the combination of the individual's genetic material that is selected in function of the good performance (objective function). To explain the multipoint crossover for each fixed number $g=1,..,n_g$, where n_g is the number of total generations, let us introduce the matrix F_g which is the set of parents for a given population. This is a matrix of dimension $F_g : n_p x N_i$, where n_p is the number of parents at the generation g and N_i is the size of every array (chromosomes). In our case, this depends of the type of radius $r \in N_r(c)$ of the CNN selected. Let $C_1(F_g, n_t)$ be the crossover operator which can be defined as the combination between the information of the parents set considering the number of intervals n_t of each individual and the number of sons n_s:

$$n_s = n_p^{\;n_t} \tag{7}$$

Then $C(F_g, n_t) : n_p x N_i \rightarrow n_s x N_i$. To show how the crossover operator can be applied, the following example is introduced. Let F_g has $n_p = 2$ and $N_i = 3$. This means that the array (the information of one father) can be divided in 3 sections and every section is determined with a_i, b_i, and c_i respectively. The number of intervals it is equivalent to the number of multipoint crossover plus one. In this case very simple case the number of intervals is 3 and the number of multipoint crossover is equal to 2.. Note that with this operator, the parents F_g of the population g are included in the result of the crossover (←). Remember, that every father corresponds to the strategy of every player.

Example:

$$F_g = \begin{bmatrix} p^1 \\ p^2 \end{bmatrix}, \begin{matrix} p^1 = \begin{bmatrix} a^1 & b^1 & c^1 \end{bmatrix} \\ p^2 = \begin{bmatrix} a^2 & b^2 & c^2 \end{bmatrix} \end{matrix}, \Rightarrow C_1(F_g) = \begin{bmatrix} a^1 & b^1 & c^1 \leftarrow \\ a^1 & b^1 & c^2 \\ a^1 & b^2 & c^1 \\ a^1 & b^2 & c^2 \\ a^2 & b^1 & c^1 \\ a^2 & b^1 & c^2 \\ a^2 & b^2 & c^1 \\ a^2 & b^2 & c^2 \leftarrow \end{bmatrix}$$

Depending of the size of templates it is possible that the number of crossover combinations is very big. In this case, it is necessary to select with some fixed probability the proportion of the population that is going to be selected.

4.2 Mutation

The mutation operator just changes some elements of every template that were selected in a random way from a fixed probability factor P_m; in other words, we just vary the components of some *genes*. This operator is extremely important, because assures the maintenance of the diversity inside the population, which is basic for the evolution. This operator $M : n_s x N_i \rightarrow n_s x N_i$ changes with probability P_m a specific population in the following way:

$$M\left(F_{ij},P_m\right)=\begin{cases}\overline{F}_{ij} & r(\omega)\leq P_m\\ F_{ij} & r(\omega)>P_m\end{cases} \tag{8}$$

$\overline{F}_{ij}\in\left[-e,e\right]\in\Re$ represents the element of the mutated strategy. Considering that the images used normally belong to discrete intervals between 0 and 255, the mutation is generated using random numbers with fixed increments that can be selected in the program. With this option it is possible to reduce the searching space of the templates and, off course, the computing time to find the optimal solution.

4.3 Selection Process

This process S_g computes the objective function O_g that represents a specific criterion to maximize or minimize, and it selects the best n_p individuals of A_g as:

$$S_g\left(A_g,n_p\right)=\max{}^{n_p} O_g\left(A_g\right) \tag{9}$$

Then, the parents of the next generation can be computed as:

$$F_{g+1}=S_g\left(A_g,n_p\right) \tag{10}$$

The objective function can be the square error or the hamming distance in the case of binary images. In this step a fixed big number is added to the objective function if the template is not stable to eliminate it from the population.

4.4 Add Random Parents

The learning of CNN is very complex because in many cases the error doesn't have the concavity condition and it is possible to stand in local minima. To avoid this problem a new scheme called adds a random parent is introduced. The procedure is the following: if the best individual of one generation or in the last *"n"* generations is the same than the previous one, new random individuals are included like parents of the next generation. This step increases the population due to the application of crossover with the new random parent. This step is tantamount to have a very big mutation probability and to search in new points of the solution space.

4.5 Results

First of all, a training set was generated using all the possible combinations of the game of life. This means 2^9=512 combinations considering the border conditions with the corresponding outputs. The parameters used for the training set are:

Table 2. Training Parameters

PARAMETER	VALUE
# of fathers	4
# of maximum random parents	3
Percentage mutation	0.15
Initial population	20000
Increment	1

Table 3. Results of the simulations

a_c	a_p	b_c	b_p	p_c	p_p	q_c	q_p	r_c	r_p	i
41	0	0	-1	1	0	-1	-4	4	-4	-1
30	0	0	-1	-2	0	-1	-2	-16	0	-1

Table 3 shows the simplest solution obtained with the algorithm. The error was computed using the Hamming Distance plus the sum of the absolute of all the terms divided by 100. With the information of these experiments, other mathematical approach was obtained and published [37]. In this paper, we proof that it is possible to find that all the solutions of the system can be represented in only three types. The solution obtained in this work with GA, corresponds to the second and third class.

5 Conclusions

Polynomial Cellular Neural Networks with only one layer can classify linearly nonseparable data points that can not be classify by a traditional CNN. This means that in hardware implementation for a real application it is necessary only one layer to solve problems that previously with CNN were solved with multilayer CNN. Moreover, with this model it is possible to perform the game of life in the same way that a cellular automata. This means that the PCNN has the same computational power of a Universal Turing Machine. With this approach it is possible to solve any problem with the same type of complexity that the game of life.

References

[1] Chua, L.O., Yang, L.: Cellular neural networks: Theory and Applications. IEEE Trans. Circuits Syst. 35, 1257–1290 (1988)

[2] Chua, L.O., Shi, B.E.: Exploiting Cellular Automata in the design of cellular neural networks for binary image processing. Technical Report No. UCB/ERL M89/130, EECS Department University of California, Berkeley (1989)

[3] Crounse, K.R., Fung, E.L., Chua, L.O.: Efficient Implementation of Neighborhood Logic for Cellular Automata via the Cellular Neural Network Universal Machine. IEEE Transactions on Circuits and Systems- I. Fundamental Theory and Applications 44(4), 355–361 (1997)

[4] Crounse, K.R., Fung, E.L., Chua, L.O.: Efficient Implementation of Neighborhood Logic for Cellular Automata via the Cellular Neural Network Universal Machine. IEEE Transactions on Circuits and Systems—I: Fundamental Theory and Applications 44(4), 355–361 (1997)

[5] Chua, L.O., Roska, T., Venetianer, P.L.: The CNN is Universal as the Turing Machine. IEEE Transactions on Circuits and Systems-I: Fundamental Theory and Applications 40(4), 289–291 (1993)

[6] Berlekamp, E., Conway, J.H., Guy, R.K.: Winning ways, ch. 25, pp. 817–850. Academic Press, New York (1982)

[7] Chua, L.O., Roska, T., Venetianer, P.L.: Zarandy, Some novel capabilities of CNN: game of life and examples ofmultipath algorithms. In: Proceedings of Second International Workshop on Cellular Neural Networks and their Applications, 1992. CNNA 1992, October 14-16, pp. 276–281 (1992)

[8] Minsky, M., Paper, S.: Perceptrons: An introduction to Computational Geometry. MIT Press, Cambridge (1969)

[9] Hassoun, M.H.: Fundamentals of Artificial Neural Networks. A Bradford Book, MIT Press (1995)

[10] Balsi, M.: Generalized CNN: Potentials of a CNN with Non-Uniform Weights. In: Proceedings of IEEE Second International Workshop on Cellular Neural Networks and their Applications (CNNA 1992), Munich, Germany, October 14-16, pp. 129–134 (1992)

[11] Bilgili, E., Göknar, I.C., Ucan, O.N.: Cellular neural network with trapezoidal activation function. Int. J. Circ. Theor. Appl. 33, 393–417 (2005)

[12] Dogaru, R., Chua, L.: Universal CNN cells. International Journal of Bifurcation and Chaos 9(1), 1–48 (1999)

[13] Gomez-Ramirez, E., Pazienza, G.E., Vilasis-Cardona, X.: Polynomial Discrete Time Cellular Neural Networks to solve the XOR problem. In: The 10th IEEE International Workshop on Cellular Neural Networks and their Applications, The Marmara Istanbul Hotel, Istanbul, Turkey, August 28–30 (2006)

[14] Chua, L., Yang, L.: Cellular neural networks: theory. IEEE Trans. Circuits Syst. 35, 1257–1272 (1988)

[15] Harrer, H., Nossek, J.: Discrete-time cellular neural networks. International Journal of Circuit Theory and Applications 20, 453–467 (1992)

[16] Roska, T., Kék, T., Nemes, L., Zarándy, Á., Szolgay, P.: CSL-CNN software library, version 7.3 (templates and algorithms) (1999)

[17] Yang, Z., Nishio, Y., Ushida, A.: Templates and algorithms for two-layer cellular neural networks neural networks. In: Proc. of IJCNN 2002, vol. 2, pp. 1946–1951 (2002)

[18] Schonmeyer, R., Feiden, D., Tetzlaff, R.: Multi-template training for image processing with cellular neural networks. In: Proc. of 2002 7th IEEE International Workshop on Cellular Neural Networks and their Applications (CNNA 2002), Frankfurt, Germany (2002)

[19] Laiho, M., Paasio, A., Kahanen, A., Halonen, K.: Realization of couplings in a polynomial type mixed-mode cnns. In: Proc. of 2002 7th IEEE International Workshop on Cellular Neural Networks and their Applications (CNNA 2002), Frankfurt, Germany (2002)

[20] Corinto, F.: Cellular Nonlinear Networks: Analysis, Design and Applications. PhD thesis, Politecnico di Torino, Turin (2005)

[21] Gómez-Ramírez, E., Pazienza, G.E., Vilasís-Cardona, X.: Polynomial discrete time cellular neural networks to solve the XOR problem. In: Proc. 10th International Workshop on Cellular Neural Networks and their Applications (CNNA 2006), Istanbul, Turkey (2006)

[22] Niederhofer, C., Tetzlaff, R.: Recent results on the prediction of EEG signals in epilepsy by discrete-time cellular neural networks DTCNN. In: Proc. IEEE International Symposium on Circuits and Systems (ISCAS 2005), pp. 5218–5221 (2005)

[23] Gómez-Ramírez, E., Najim, K., Ikonen, E.: Forecasting Time Series with a New Architecture for Polynomial Artificial Neural Network. Applied Soft Computing 7(4), 1209–1216 (2007)

[24] Berlekamp, E., Conway, J.H., Guy, R.K.: Winning ways for your mathematical plays. Academic Press, New York (1982)

[25] Rendell, P.: A Turing machine in Conway's game life (2006), `http://www.cs.ualberta.ca/~bulitko/f02/papers/tmwords.pdf`

[26] Pazienza, G.E., Gomez-Ramirez, E., Vilasis-Cardona, X.: Polynomial Cellular Neural Networks for Implementing Semitotalistic Cellular Automata. In: NNAM 2007, Neural Networks and Associative Memories, Magno Congreso Internacional de Computación, Instituto Politécnico Nacional, IPN; México, 5 al 9 de Noviembre de (2007)

[27] Matsumoto, T., Chua, L., Furukawa, R.: CNN cloning template: connected component detector. IEEE Trans. Circuits Syst. 37(5), 633–635 (1990)

[28] Zarándy, Á.: The art of CNN template design. International Journal of Circuit Theory and Applications 27, 5–23 (1992)

[29] Magnussen, H.:: Discrete-time cellular neural networks: theory and global learning algorithms. PhD thesis, Techinical university of Munich, Munich (1994)

[30] Kozek, T., Roska, T., Chua, L.: Genetic algorithm for CNN template learning. IEEE Trans. Circuits Syst. 40(6), 392–402 (1993)

[31] Magnussen, H., Nossek, J.: Global learning algorithms for discrete-time cellular neural networks. In: Proc. third IEEE International Workshop on Cellular Neural Networks and their Applications (CNNA 1994), Rome, Italy, pp. 165–170 (1994)

[32] Kellner, A., Magnussen, H., Nossek, J.: Texture classification, texture segmentation and text segmentation with discrete-time cellular neural networks. In: Proc. third IEEE International Workshop on Cellular Neural Networks and their Applications (CNNA 1994), Rome, Italy, pp. 243–248 (1994)

[33] Alander, J.T.: An Indexed Bibliography of Genetic Algorithms: Years 1957–1993 (1994); Art of CAD ltd

[34] Bedner, I.: Genetic Algorithms and Genetic Programming at Stanford, Stanford Bookstore (1997)

[35] Holland, J.H.: Adaptation in Natural and Artificial Systems. MIT Press /Bradford Books Edition, Cambridge (1992)

[36] Gómez-Ramírez, E., Mazzanti, F.: Cellular Neural Networks Learning using Genetic Algorithm, 7mo. Congreso Iberoamericano de Reconocimiento de Patrones. In: CIARP 2002, CIC-IPN, Cd. de México, México, Noviembre 19-22 (2002)

[37] Pazienza, G.E., Vilasis-Cardona, X., Gomez-Ramirez, E.: Polynomial Cellular Neural Networks for Implementing the Game of Life. In: de Sá, J.M., Alexandre, L.A., Duch, W., Mandic, D.P. (eds.) ICANN 2007. LNCS, vol. 4668, pp. 914–923. Springer, Heidelberg (2007)

EMG Hand Burst Activity Detection Study Based on Hard and Soft Thresholding

Mario I. Chacon Murguia[1], Leonardo Valencia Olvera[2], and Alfonso Delgado Reyes[3]

[1] Chihuahua Institute of Technology
[2] Nogales Institute of Technology
[3] Chihuahua Autonomous University, Mexico

1 Introduction

Electric signal analysis from live organism is an old area that was documented by Francesco Redi dated from 1666, Walsh 1773, and Galvani 1792 [1]. Contraction of muscular fibers by electric impulses was recorded by Debois-Raymmod 1849 [1]. Electric impulses known as myolectric signal and their recording are named electromyographic signals or EMG [2-8]. The first clinical use of EMG signals was reported in 1966 by Harddyck. It is not until the 1980´s that clinical methods to monitor EMG of several muscles were achieved [1].

Notwithstanding the ancient origin of EMG analysis field, it is currently a promising area. Nowadays EMG research involves, human body movement studies [9], and robotic prostheses [3,10]. Still other EMG applications are related to neurological and neuromuscular problem diagnosis, and therapy [10]. Despite the many years involved in EMG research there is still specific aspects on EMG signals analysis that require more research. One of these important aspects is the determination of the starting (onset) and ending (offset) times of muscle activity termed burst detection. Burst detection is a complex problem because of the high level of noise present in the EMG signals, and the burst variability due to physical characteristics of the individuals.

Recent advances in signal processing techniques have made possible to develop several techniques for burst detection. Some of the proposed techniques include, Fourier and Wavelet transform [1,7,8] time – frequency analysis [1], Wigneer–Ville distribution [1], statistic metrics, and rank order statistics [11], even new emergent techniques based on computational intelligence has been reported [1]. Although many methods for burst detection have been proposed, it is an open problem that requires new proposals to achieve more robust methods.

The objective of the research described here deals with the study of four burst detection methods, including the proposal of two new methodologies, one based on cumulative difference sums, and other in clustering. In addition, a proposal of a new metric to determine the performance of burst detection algorithms is described.

P. Melin et al. (Eds.): Bio-inspired Hybrid Intelligent Systems, SCI 256, pp. 177–195.
springerlink.com

The work is organized as follows. Section 2 presents the information relevant to EMG signals and their acquisition. Current detection methods reported in the literature are analyzed in Section 3. The proposed preprocessing methodology of the EMG signals is described in Section 4 and the detection methodologies are covered in section 5. Section 6 explains the development of the burst detection performance index and presents the evaluation of the burst detection methods. The results and conclusions of the research are presented in Section 7.

2 EMG Signals

2.1 EMG Signal

A surface electromyographic signal SEMG consist of action potential trains, AP´s, generated by a number of motor units, MU, and can be represented by

$$s(n)=\sum_{r=0}^{N-1} h(r)\varphi(n-r)+\eta_G(n) \tag{1}$$

where $\varphi(n)$ is the firing impulse, $h(r)$ represents the motor unit action potentials, MUAP, $\eta_G(n)$ is additive Gaussian noise of zero mean, and N is the number of firing impulses of the MU.

2.2 Signal Acquisition

The signals used in this research were acquired with the MP-35 system from BIOPAC *Systems*® and the software BSL PRO 3.7. The electrodes used were surface electrodes from the series EL503 and EL507.

The signals come from the following hand movements:

Hand at rest, 1 recording.
Hand opening and closing, 1 recording.
Hand at gradual prehensile movement from minimum to maximum, 5 recordings.
Hand at four regulated prehensile movement, 5 recordings.
Hand at four gradual increasing prehensile movement, 5 recordings.

These signals were obtained from five individuals that are identified as M, F, C, Y, and S for documentation purposes.

3 Current EMG Burst Detection Methods

From a literature review we can say that most of the methods relay on the definition of a threshold to decide muscle activity. The methods found are based on statistic, wavelet, nonlinear operators, and rank order statistics.

3.1 Statistical Approach

The simplest statistic method is proposed by Abbink et al [6]. The threshold is obtained from the mean μ and standard deviation σ of the histogram of samples of muscle activity signals acquired from a defined rhythmic movement. This histogram approximates a normal distribution. The threshold is defined by

$$Th_{EMG} = \mu + \gamma\sigma \tag{2}$$

where γ is the parameter to establish the distance of the threshold with respect the mean value. According to the literature it is set to values in the range of 1 to 3 and it is determined experimentally.

3.2 Wavelet

Merlo et al, [7], and Gazzoni et al [8] proposed a wavelet based method. The method tries to identify the MUAP by wavelet analysis. This approach considers the a priori physiologic knowledge of the EMG signal and the individual MUAPs components. Muscle activity is detected based on the presence of MUAPs in the EMG. Merlo et al [8] established a simplified model of the EMG signal defined as the motor unit action potential train, MUAPT formulated as

$$MUAPT_j(t) = \sum_i \beta_j f\left(\frac{t-\theta_{ij}}{\alpha_j}\right) \tag{3}$$

where j is a specific MU, β_j is the amplitude factor, θ_{ij} is the time occurrence of the MUAPT of the MU, and α_j is the scale factor. *f()* is assumed to be the function that approximate a generic MUAP detected on the skin surface at time t. The SEMG including the noise factor, *n(t)*, can be represented by

$$s(t) = \sum_i MUAPT_i(t) + \eta(t) = \sum_i \sum_j \beta_j f\left(\frac{t-\theta_{ij}}{\alpha_j}\right) + \eta(t) \tag{4}$$

In this way the SEMG signal can be transformed to the time-scale domain by the wavelet transform

$$CWT(a,\tau) = \frac{1}{\sqrt{a}} \int_{-\infty}^{+\infty} s(t) W\left(\frac{t-\tau}{a}\right) dt \tag{5}$$

where *w()* is the wavelet prototype, τ is the translation index, and a is the scale factor.

3.3 Nonlinear Operators

This operator dates from the work of Teager in 1983, and in 1990 Kaiser developed the TKEO for the discrete domain [12,13,14] which is defined as

$$\Psi\left[x(n)\right] = x^2(n) - x(n+1)x(n-1) \tag{6}$$

The application of the TKEO operator to detect muscle activity in SEMG signals is reported in [13]. Results of this method show a dramatically improvement of the SNR which increases the probability to detect the burst activation times, especially in low SNR. The application of the TKEO in SEMG signals enhances the activity of motor units. Lin and Aruin in [13], presented a burst detection method based on the energy operator Teager-Kaiser, TKEO. Once the TKEO is applied, the resulting signal is rectified and a threshold is obtained from it based on the mean and standard deviation to detect the bursts.

The TKEO is a nonlinear operator. Its output depends on the amplitude and frequency product of the signal, and has been applied in different areas like signal demodulation, and electrophysical signals. Other applications are related to time-frequency distribution generation in signals [13].The TKEO operator has also been used in speech processing, AM–FM demodulation, detection of transitory events that contaminate AM and FM signals, neural spike detection, as well as in encephalogram signals. Other applications can be found in [14]. The TKEO differentiate the energies of the low changing modulated signals in amplitude and in frequency from short duration and fast changing transitory events of amplitude and frequency [l].

The idea of using the TKEO to detect EMG burst, comes from this previous application. The situation presented in [14] has some similitude with the problem that we treat in this work related to discriminate differences of amplitude, frequency and duration between muscular activity and background noise. Therefore the TKEO can be used to differentiate the muscle activity event from the non-activity event [12]. There exist some restrictions on the application of the TKEO related to broad frequency range signals and very noisy background. However, the TKEO can be used in our work because our signal frequencies are located in the range of 5-500Hz.

3.4 Rank Order Statistic Filters

The EMG surface signal can be considered as a Gaussian process $s(t)=N(0,\sigma)$ modulated by the muscular activity and additive Gaussian noise $n(t)=N(0,\sigma_n)$. The latter due to thousands of muscular cells that are activated at the same time. Under this consideration Rank Order Filters have also been proved to obtain the firing burst, [9].

4 EMG Processing and Filtering

The works described in the previous section depend only on one technique missing the advantages provided by the other techniques. Therefore, it was decided to develop a new methodology based on information fusion of three techniques in order to take advantage of the information provided for them. The techniques involved are; non linear energy operator Teager–Kaiser, Wavelet, and conventional filtering.

4.1 DC Removal and Full Wave Rectification

The first processing of the EMG consists on the removal of the DC component. This process is achieved by eliminating the mean or lineal tendency presented in the EMG signal, due to individual physiology or due to electrical induction during data acquisition. The process consist on computing the straight line that fits the data obtained with the least-squares and then subtracts the resulting function from the data. Next, the EMG signal is rectified using the absolute value. This process is applied because we are only interested on burst detection, and it allows us to use only positive values that simplify other processing.

4.2 TKEO Application Results

The results of the application of the TKEO method are illustrated for the individual cases A, F, M, S and Y, in Figure 1 a) to e) respectively. The sequence in each plot corresponds to the original EMG signal, and the rectified signal. It is easy no notice the burst activity enhancement after the processing.

4.3 Wavelet Filtering

In order to obtain the best results in the detection of the bursts presented in an EMG, it is necessary to find the best match between the wavelet function and the burst. The first order Hermite-Rodriguez function,_$HR_1(t)$, has been used in the case of simple differential recordings to describe the basic form of the MUAP [7]. In our case, the recordings were obtained using a double differential electrode; therefore it is recommended to select the wavelet function accordingly to this modality. The scale parameter a is selected such that the wavelet duration is in the range of the physiological duration of the MUAP, (5 to 40 ms). The characteristics of the MUAP were used to select the wavelet function, which are similar to the basic form of the wavelets *db2* and *db4*. Since we only look for an approximation of the form of the wavelets *db2* and *db4*. Since we only look for an approximation of the burst the requirements are not as strong as reported in [1]. According to [1] where a unipolar recorded signal is captured it is recommended to approximate the MUAP by the second derivative of a Gaussian function. This work suggested the use of the *Mexican Hat* wavelet. The drawback of this wavelet is that the "*Mexican Hat*" function does not match with the MUAP shape. Pattichis and Pattichis documented in [1] stated that by using wavelet multiresolution analysis, the coefficients of each stage could be used to build an approximation of the original signal. Kumar et al cited in [1], also established that it is possible to determine the muscular fatigue in surface EMG by performing wavelet decomposition with *Sym4* or *Sym5* in levels 8 and 9. Using this information, it is possible to do a sampling analysis of the CWT using only a few scales and we can obtain the outputs of the filter bank with a specific degree of correspondence with the MUAP selected.

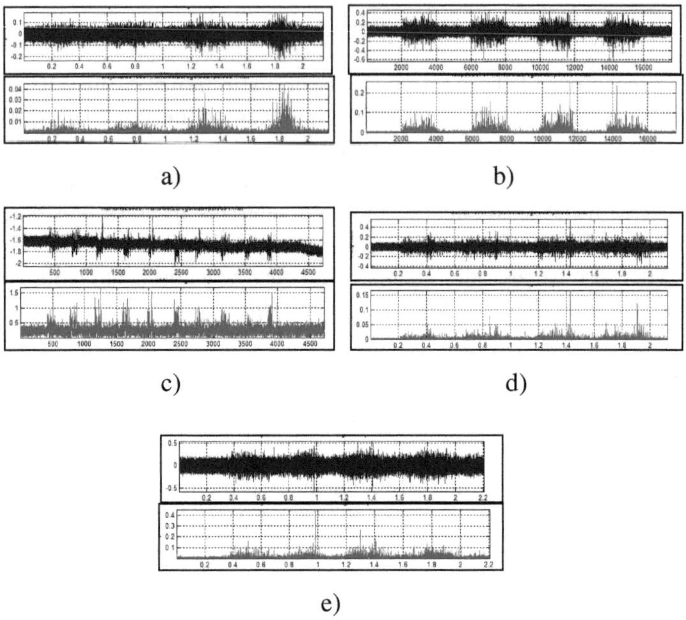

Fig. 1. TKEO operator application to the EMG signals. a) Individual A, b) Individual F, c) Individual M, d) Individual S, e) Individual Y.

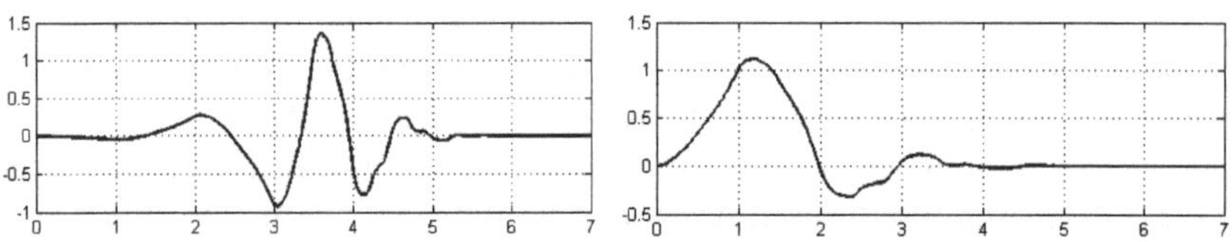

Fig. 2. *db4* wavelet and scale functions.

Looking for a better similitude between a bioelectric event and the wavelet function, we found that the wavelet and scale functions *db4* are the more appropriated, Figure 2. It can be noticed that this function has a smoother transition than the previous functions, and it better matches the subpolarization (depolarization) and overpolarization (hyperpolarization) of the muscular fibers, thus representing a better choice.

4.3.1 Wavelet Analysis Process

The Wavelet analysis is achieved with the *db4* wavelet, the threshold selection criterion used is the *minmax*, the thresholding algorithm applied is the *Hard Thresholding*, and the multiplicative scaling factor of the threshold selected was *sln* (terms

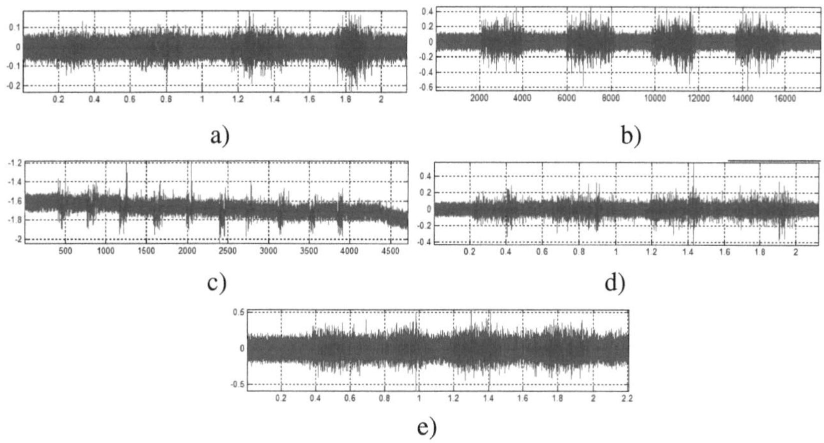

Fig. 3. Wavelet analysis for EMG denoising. Individuals a) A, b) F, c) M, d) S, and e) Y.

according to Matlab terminology). The decomposition is achieved in level 5 and the wavelet is *db4*. The noise reduction is considerable and the burst is well defined, as illustrated in Figure 3. Another possibility of the thresholding method may be the *Soft Thresholding*, however, for the case of EMG signals the *"Soft Thresholding"* presented more amplitude attenuation than the *"Hard Thresholding"*.

The case of the individual A shows high noise reduction in non-activity portions, and well-located burst. We also have good noise reduction rate and better burst definition for individual F. The EMG signal of the individual M presents a DC tendency component, however, noise reduction, allows burst definition. The DC tendency is corrected in other stage, making possible a better burst definition. Albeit the signal has a high noise component for the individual S the wavelet analysis method yields an adequate reconstruction. Finally, for individual Y the EMG bursts are barely perceived in the original signal, nevertheless, the wavelet process is able to extract the burst information. In conclusion, we can say that the proposed wavelet analysis can reduce the noise component presented in the EMG signal and make it possible to better define the burst.

4.4 Butterworth Filtering

The Butterworth filter design is achieved considering that the MUAP has a duration between 5ms and 40 ms [1], which corresponds to 25Hz to 200Hz. The EMG acquisition process was performed in a 5Hz to 500Hz band with a sampling frequency of 1000 samples/sec. In [1] the EMG envelope is obtained with a low pass 4th order Butterworth filter with a cutoff frequency of 10Hz, and using face compensation. In this work the low frequency components lower than 25Hz are first eliminated with a high pass filter. Then a low pass filter is applied to eliminate frequencies higher than 200 Hz. A new low pass filter with cutoff frequency of 10 HZ is applied to obtain the EMG envelope.

5 Burst Detection Methods

Once the EMG signal has been enhanced, the final problem to solve is to determine the starting (onset) and ending (offset) points of each burst present in the EMG signal. The most dominant solution reported in the literature [1,3], and [7] indicates that this problem is solved by setting an activation threshold level. The method proposed in this work to detect the burst activity includes semiautomatic and automatic approaches.

In the semiautomatic detection method the user only has to define the activation threshold level that will be used in the detection process. This method includes the conventional statistical technique and the differences of the values of the cumulative sum differences of the signal sample amplitudes. This approach is based on two threshold determination techniques; empirically determined from the differences of the values of the cumulative sum differences of the signal sample amplitudes, DSA, and the simple statistics, ES.

The activation threshold level is automatically determined in two other methods. The approaches considered here are: establishing the threshold based on the centroid that represents the lower amplitude values of the EMG signal, obtained by the Fuzzy C-Means clustering algorithm, FCM [15], and that obtained by the K-Mean algorithm KM [16].

The experiments reported here were performed using five individuals considering the movements specified in Section 2.3.

5.1 Semi-automatic Detection

In this method the threshold was set using sections of the EMG signal without activity. This type of signal is unique for each individual and for each muscle. The non-activity signal provides important information like possible noise present during an activity signal.

5.1.1 Statistical Threshold

The noise present during the rest state is characterized by its mean, μ_n, and standard deviation, σ_n. These statistics allow us to establish a threshold parameter γ that represents the maximum amplitude reached by the noise, equation (7).

$$Th_{EMG} = \mu_n + \gamma\sigma_n \tag{7}$$

In this way, we can determine what sections of the signal corresponds to noise and what to a burst. μ_n and σ_n are used to define the variations of the threshold offset during the experimentation. The threshold has a lower limit equal to μ_n, and it can linearly increase with respect to the number of σ_n accepted. The literature

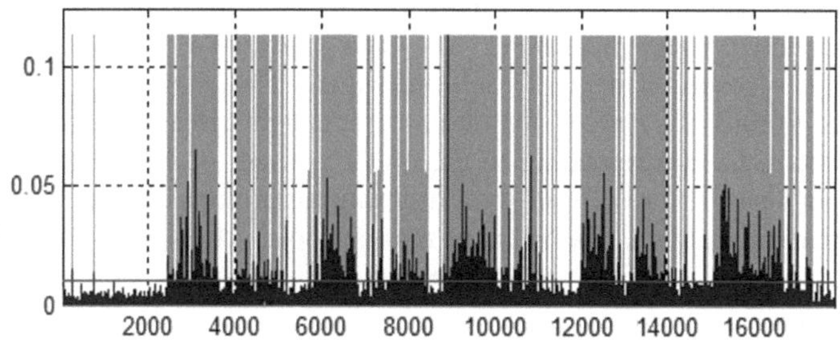

Fig. 4. Burst detection using the statistical threshold.

recommends [6] that an acceptable range of γ is between 1 and 3, however it depends on the natural individual variation. Another possibility is to consider the noise as a Gaussian process with zero mean, therefore it is possible to set γ=3, which warranties the 99% of the noise will be cover by the threshold. Figure 4 illustrates the statistical threshold, indicated in red, the blue trace corresponds to the EMG signal, and the samples greater than the threshold are in green.

5.1.2 Cumulative Sum Differences

The second semiautomatic method is based on the cumulative sum differences *DSA*. The fundament of the method is the representation of the envelope or burst description. There is a cumulative sum of the envelope samples, and a difference between points separated at certain distance is computed. This difference is compared against a threshold, determined empirically. If the difference is greater than the threshold it indicates that the signal is still evolving inside of a burst, and a trigger is generated. These triggers indicated the detection signal. Figure 5 shows an example of the cumulative sum differences and the signal that corresponds to detected regions of the bursts.

5.2 Automatic Detection

Automatic detection refers to the method where the activation threshold is obtained by data analysis without human intervention. The methods used to determine the activation thresholds are the Fuzzy C-means, and K-means clustering

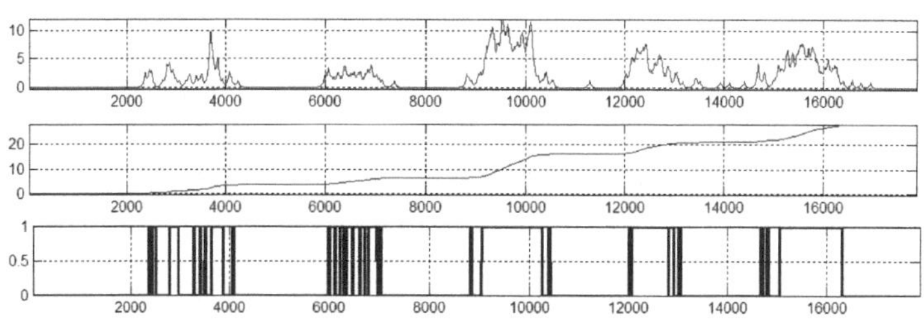

Fig. 5. DSA method applied to an EMG envelope.

algorithms. These methods were used to determine the activation thresholds for each individual for each channel. The data analyzed is one recording of each movement of each channel of each individual. The cluster algorithms are programmed to find two classes, the onset class, activation, and the offset class, rest. The centroid, C, corresponding to the offset is used to classify the samples into activity or rest. The results of the clustering algorithms for each individual are summarized in Table 1. The value of the threshold, Th, is obtained from the centroid value plus a 20% of its value. This criterion was defined by posterior experimentation.

6 Performance Analysis

In general, the performance of burst detection methods in EMG signal is achieved by human evaluators. In order to obtain an automatic performance evaluation and reduce possible subjective factors on the evaluation we decided to develop a performance index based on quantitative measurements. The performance index developed here considers as the reference the results of the manual detection method. This criterion was selected in order to compare the proposed methods performances against the human criterion.

The burst patterns to be compared with the other methods are obtained as follows. The signal presented to the user to detect the burst is the signal processed by, rectification, TKEO, denoised Wavelet approximation, and the signal envelope generated by Wavelet-Butterworth. From this signal the expert marks the starting and ending points of the burst in each EMG signal. Figure 6 shows the four signals, EMG rectified, TKEO, wavelet denoised, and the envelope obtained with the Butterworth filter. The starting and ending points are defined in the last signal, and translated to the others.

Table # 1. Centroid values for each individual

Individual		A		F		M		S		Y	
Chanel		1	2	1	2	1	2	1	2	1	2
KM	C	0.0008	0.0014	0.0019	0.0031	0.0016	0.0038	0.0004	0.0014	0.0016	0.0035
	Th	0.0009	0.0017	0.0022	0.0037	0.0019	0.0045	0.0005	0.0017	0.0019	0.0042
FCM	C	0.0006	0.0013	0.0013	0.0024	0.0015	0.0032	0.0004	0.0012	0.0014	0.003
	Th	0.0008	0.0015	0.0016	0.0028	0.0018	0.0038	0.0004	0.0014	0.0017	0.0037

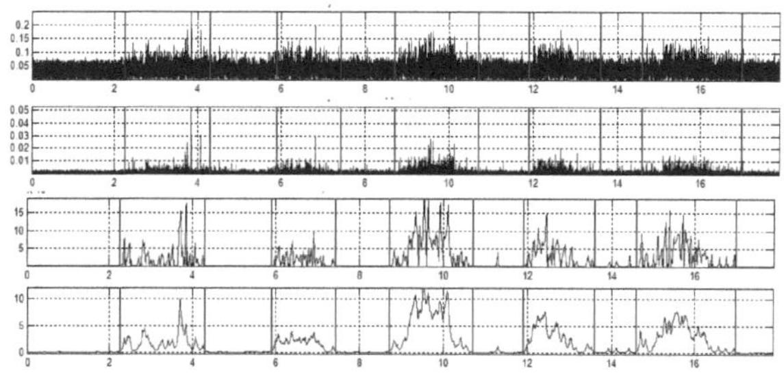

Fig. 6. Burst patterns generation. a) EMG Rectified, b) TKEO rectified, c) wavelet denoised, d) Wavelet-Butterworth.

The proposed performance index and its application to the proposed methods are described next.

6.1 Performance Index

Let an EMG signal be represented by $\boldsymbol{X}$ and let $\boldsymbol{X_M}$ representing its manual burst detection. Now assume $\boldsymbol{X_A}$, as the automatic bursts detection of $\boldsymbol{X}$ by an automatic method $\boldsymbol{A}$. Then the starting and ending points of each burst in $\boldsymbol{X_M}$ and in $\boldsymbol{X_A}$ are determined by generating the vectors $\mathbf{V}_{xM}$ and $\mathbf{V}_{xA}$ respectively. The performance of the method $\boldsymbol{A}$ is computed by comparing the vectors $\mathbf{V}_{xM}$ and $\mathbf{V}_{xA}$. The result of this comparison will be called Burst Correspondence Index, *BCI*. The *BCI* indicates a measure of matching among the detected bursts in both vectors $\mathbf{V}_{xM}$ and $\mathbf{V}_{xA}$ yielding a performance index for the method $\boldsymbol{A}$. Each burst in $\mathbf{V}_{xM}$, B_{xM}, and in $\mathbf{V}_{xA}$, B_{xA} is represented by its starting and ending points denoted by $\{S_{BX_M}, E_{BX_M}\}$ and $\{S_{BX_A}, E_{BX_A}\}$. The *BCI* is composed of two metrics:

i) Ratio of the length or duration of the burst, R_{LB} .
ii) Burst starting and ending points matching, S_B

The ratio of the length or duration of the burst, R_{LB} is formulated as

$$R_{LB} = \frac{L_{BX_A}}{L_{BX_M}} \qquad 0 \le R_{LB} \le 1 \tag{8}$$

where

$$L_{BX_M} = E_{BX_M} - S_{BX_M} \qquad L_{BX_M} = E_{BX_A} - S_{BX_A} \tag{9}$$

The burst starting and ending points matching, S_B, tells us how precise was the detection of the burst in time, and it is defined by

$$S_B = \frac{L_{BX_A}}{L_{BX_M} + (\Delta_{SB} + \Delta_{EB})} \tag{10}$$

where

$$\Delta_{SB} = \left|S_{BX_A} - S_{BX_M}\right| \qquad \Delta_{EB} = \left|E_{BX_A} - E_{BX_M}\right| \tag{11}$$

represents the distance among the starting and ending points of the burst. Equation (11) is used in equation (10) to penalize the precision in time of the burst detection. Equation (11) is a simple absolute measure. However, more sophisticated measures can be derived if the sign of the differences is considered. In this case, information related to advance (overshoot) or lag in burst detection may be incorporated in the performance index.

The application of the proposed *BCI* is illustrated in the next case, assuming the bursts shown in Figure 7.

In this case

$$S_{BX_M} = 3000 samples,\ E_{BX_M} = 7000 samples,\ S_{BX_A} = 2500 samples, E_{BX_A} = 6500 samples$$

then

$$R_{LB} = \frac{L_{BX_A}}{L_{BX_M}} = 1 \quad \Delta_{SB} = 500, \quad \Delta_{EB} = 500 \quad \text{and} \quad S_B = \frac{L_{RX_A}}{L_{RX_M} + (\Delta_{SR} + \Delta_{ER})} = 0.8$$

Considering only the length of both burst we have a good match, however considering the time of occurrence the performance decreases to 0.8. The previous metrics allow us to measure how good the burst detection was. A perfect match in the detection of the bursts in the EMG signal should indicate that all the burst in the EMG signal has been detected with exact time precision. The *BCI* of the automatic method must be compared with a perfect match, base value, obtained from the manual vector. The base value is obtained by assuming a one to one correspondence of the burst in the manual detection vector and the automatic detection vector. We sum up all the perfect matching of the burst in the manual detection vector. That is

$$D_{BCI} = \sum_{i=1}^{N} S_{BP_i} \tag{12}$$

where N is the total number of burst in $\mathbf{V}_{BM}$ and S_{BP_i} is the performance index of one burst in the automatic vector. A perfect match of a burst corresponds to a value of $S_{BP_i} = 1$, if we have N bursts then $D_{BCI} = N$.

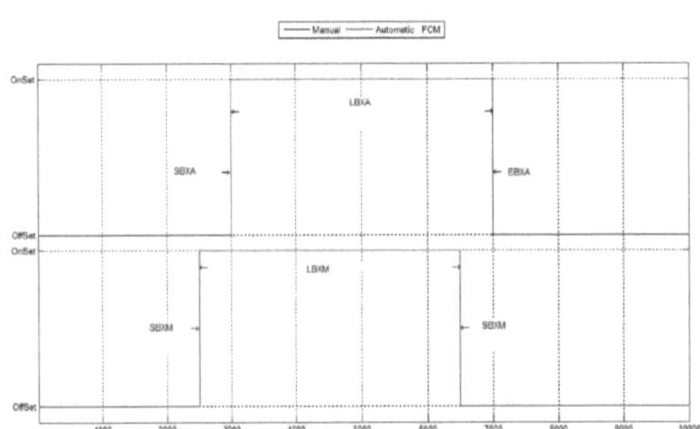

Fig. 7. Parameter determination to compute the BCI.

Once the base value is computed, we proceed to obtain the cumulative *BCI* of the automatic vector. This is achieved by summing up all the matching indexes of the burst in the automatic vector. The cumulative *BCI* for the automatic vector is given by

$$S_{TBA} = \sum_{k=1}^{M} S_{Bk} \tag{13}$$

where s_{Bk} is the *BCI* of the burst *k* and *M* is the number of burst of the automatic detection vector with correspondence to the manual detection vector, with range $1 \leq M \leq N$.

During the performance index computation some possible situations may arise. The value of *M* cannot be less than 1, because if it were 0, then it would correspond to the case where the automatic vector does not have any correspondence with the manual vector. This will imply that the manual detection performance is null or zero. This case corresponds to the worst case. Another situation is when the value of *M* is less or equal to the number of burst of the manual detection vector. If it is less, it indicates that the method detected less bursts than expected. If it is equal, it indicates the automatic method detected all burst detected in the manual detection, best case. Base on these possibilities the comparison between the base value and the similitude index of the automatic method is

$$D_{TVA} = \frac{S_{TBA}}{D_{BCI}} \tag{14}$$

where D_{VTA} is the detection performance for the automatic vector. The value of performance is $0 \leq D_{TVA} \leq 1$. The value of zero corresponds to a null detection, and the value of one corresponds to a perfect detection. For illustration consider the case shown in Figure 8. The burst of the manual vector correspond one to one to the burst of the automatic vector, that is $N = M = 4$. The *BCI* of each burst S_{Ri} is first obtained to compute the D_{TVA}.

$$D_{TVA} = \frac{S_{TBA}}{D_{BC}} = \frac{\sum_{k=1}^{M} S_{Bk}}{N} = \frac{\sum_{k=1}^{4} S_{Bk}}{4} = \frac{S_{B1} + S_{B2} + S_{B3} + S_{B4}}{4} \tag{15}$$

Assuming the *BCI´s* for each burst of the automatic detection vector to be
$S_{B1} = 0.85,\ S_{B2} = 0.93,\ S_{B3} = 0.87,\ S_{B4} = 0.96$

then $D_{TVA} = \dfrac{0.85 + 0.93 + 0.87 + 0.96}{4} = 0.9025$

Once the performance index has been established, we can proceed to evaluate the different detection methods. For the sake of simplicity of the performance, we will show graphically the performance over all the signals in separated sets, channel (dorsal), and channel 2 (ventral).

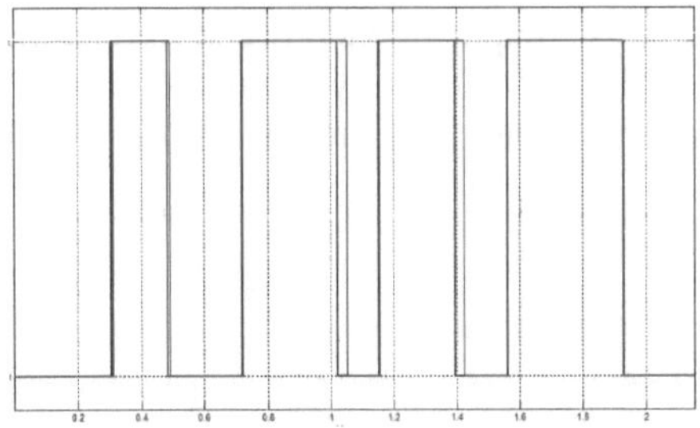

Fig. 8. Overlapping burst vectors of a case of manual and automatic detection, best matching.

6.2 Evaluation of the Proposed Methods

6.2.1 DSA Performance

Figure 9 shows the performance index for each signal in channel 1 and 2 for the method DSA. From Figure 9 we can notice that the method has better performance for channel 2 in the first 17 signals of individual A. The method has high performance in the second set of 17 signals, individual F. The next 17 signals correspond to the individual M. The final two sets of recordings correspond to persons Y and S which present some consistency. The quantitative performance of the DSA method expressed by number of signals and percentage of signals by performance ranks is shown in Table 2. Table 2 indicates the percentage of signals that falls in different *BCI* intervals, starting at 0.6. The experiment shows that for both channels there are more than 50% of the signals with $D_{TVA} \geq 0.9$.

6.2.2 FCM Performance

The results for the FCM method are shown ion Figure 10 and Table 3. For this method more than 65% of the signal in channel 1has a *BCI* greater than 0.9, and 61% for the channel 2.

6.2.3 KM Performance

The results for the KM method are shown ion Figure 11 and Table 4. This method presents a similar result for channel 1 but not for channel 2.

6.2.4 ES

The results for the ES method are shown in Figure 12 and Table 5. Results yielded by this method were not as good as the other methods. Considering the $D_{TVA} \geq 0.9$ case only 52% of the signal fall in this range for both channels.

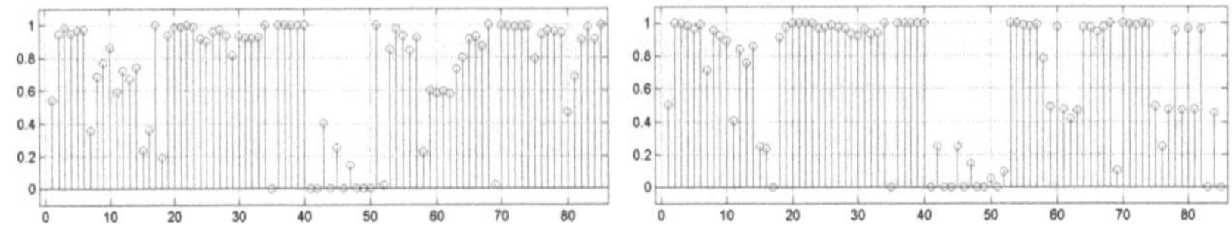

Fig. 9. DSA method performance. Channel 1 and 2.

Table 2. DSA method performance expressed by number of signals and percentage of signals by performance ranks.

Method	DSA			
Channel	CH1		CH2	
Rank	Number of signals	% Signal group	Number of signals	% Signal group
$D_{TVA} < 0.6$	24	28.24	31	36.47
$D_{TVA} \geq 0.6$	61	71.76	54	63.53
$D_{TVA} \geq 0.8$	52	61.18	51	60
$D_{TVA} \geq 0.9$	46	54.12	48	56.47
$D_{TVA} = 1.0$	11	12.94	17	20
Total Recordings	85	100	85	100

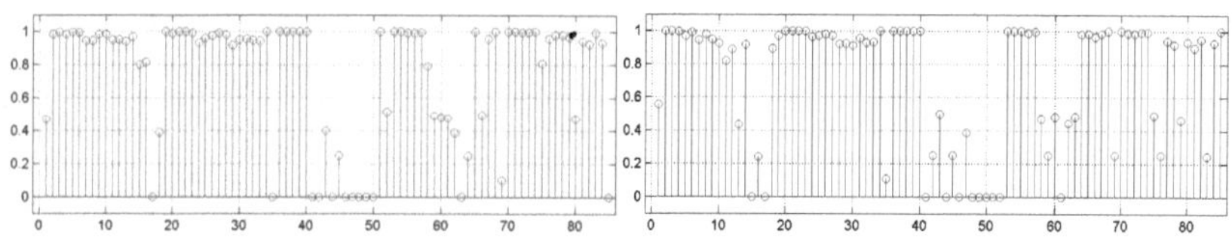

Fig. 10. FCM method performance. Channel 1 and 2.

Table 3. FCM method performance expressed by number of signals and percentage of signals by performance ranks.

Method	FCM			
Channel	CH1		CH2	
Rank	Number of signals	% Signal group	Number of signals	% Signal group
$D_{TVA} < 0.6$	25	29.41	29	34.12
$D_{TVA} \geq 0.6$	60	70.59	56	65.88
$D_{TVA} \geq 0.8$	59	69.41	56	65.88
$D_{TVA} \geq 0.9$	56	65.88	52	61.18
$D_{TVA} = 1.0$	15	17.65	18	21.18
Total Recordings	85	100	85	100

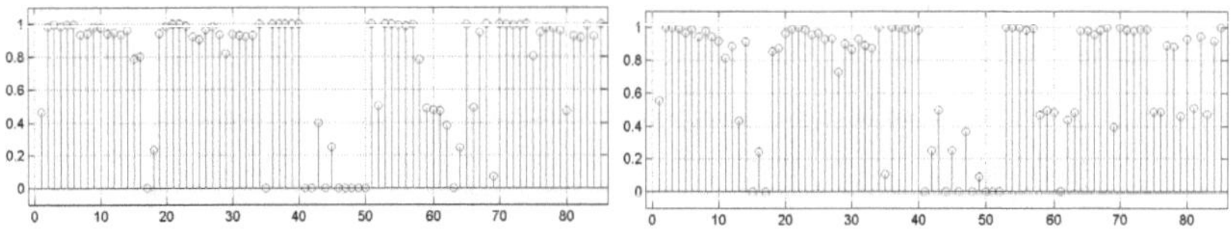

Fig. 11. KM method performance. Channel 1 and 2.

Table 4. KM method performance expressed by number of signals and percentage of signals by performance ranks.

Method	KM			
Channel	CH1		CH2	
Rank	Number of signals	% Signal group	Number of signals	% Signal group
$D_{TVA} < 0.6$	24	28.24	30	35.29
$D_{TVA} \geq 0.6$	61	71.76	55	64.71
$D_{TVA} \geq 0.8$	58	68.24	54	63.53
$D_{TVA} \geq 0.9$	56	65.88	45	52.94
$D_{TVA} = 1.0$	14	16.47	12	14.12
Total Recordings	85	100	85	100

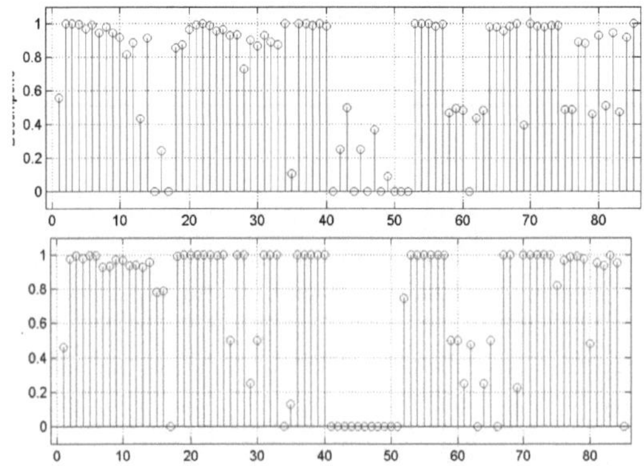

Fig. 12. KM method performance. Channel 1 and 2.

Table 5. KM method performance expressed by number of signals and percentage of signals by performance ranks.

Method	ES			
Channel	CH1		CH2	
Rank	Number of signals	% Signal group	Number of signals	% Signal group
$D_{TVA} < 0.6$	29	34.12	32	37.65
$D_{TVA} \geq 0.6$	56	65.88	53	62.35
$D_{TVA} \geq 0.8$	53	62.35	46	54.12
$D_{TVA} \geq 0.9$	52	61.18	45	52.94
$D_{TVA} = 1.0$	28	32.94	15	17.65
Total Recordings	85	100	85	100

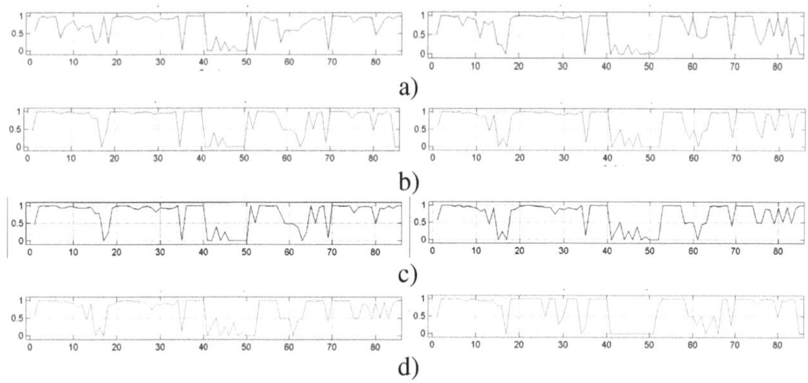

Fig. 13. Overall method performance, a)DSA, b)FCM, c)KM, d)ES, channels 1 and 2.

A summary of the performance of the methods is shown in Figure 13 and Table 6. It can be observed from Table 6 that the FCM method presents the best performance in the range $D_{TVA} \geq 0.8$ that represent an acceptable performance, followed by the KM. Also the FCM method is the most stable method considering the evaluation by range. That is, the FCM method is the better method in four out of the 5 ranges. Using the same criteria, the second best method is the KM, the third best is ES, and the last is the DSA.

Finally, the performance with respect to stability, mean, variability, and standard deviation of the four methods are shown in Figure 14. It is observed that the FCM has the best performance with respect the mean, and the KM has the best performance with respect the standard deviation.

Table 6. Overall method performance

Method	DSA	FCM	KM	ES
Chanel	CH1 & CH2	CH1& CH2	CH1& CH2	CH1& CH2
Rank	% Signal group	% Signal group	% Signal group	% Signal group
$D_{TVA} < 0.6$	32.35 (3)	31.76 (1)	31.76 (1)	35.88 (2)
$D_{TVA} \geq 0.6$	67.65 (1)	68.24 (1)	68.24 (1)	64.12 (3)
$D_{TVA} \geq 0.8$	60.59 (3)	67.65 (1)	65.88 (2)	58.24 (4)
$D_{TVA} \geq 0.9$	55.29 (4)	63.53 (1)	59.41 (2)	57.06 (3)
$D_{TVA} = 1.0$	16.47 (3)	19.41 (2)	15.29(4)	25.29 (1)

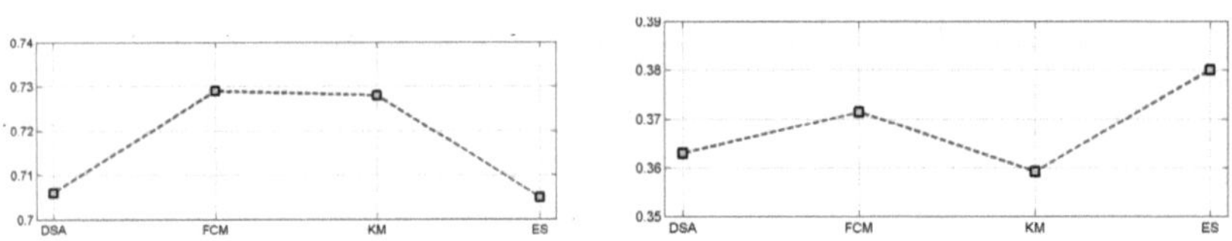

Fig. 14. Performance a) stability and b) variability of the four methods

7 Results and Conclusions

In conclusion, the results obtained in this research indicate that it is possible to automatically determine a threshold to discriminate EMG burst activity. The automatic proposed techniques based on clustering, FCM and K-mean also work better than the simple statistic and cumulative sum differences approache. The FCM and KM algorithms were the more stable methods for burst detection. From these two techniques the FCM was the most dominant according to its performance results. From the two clustering methods the best one turned out to be the FCM considering the performance evaluation with the *BCI* index, and the variability performance analysis. Our study also indicates that the clustering approaches might be a better option to define more adequate activation thresholds than statistical methods.

The proposed performance metric developed in this work is an innovative contribution in the field of EMG signal analysis. This metric quantitative and it provides objective performance evaluation of burst detection, instead that only subjective ones based on human estimation.

References

[1] Reaz, M.B.I., Hussain, M.S., Mohd-Yasin, F.: Techniques of EMG Signal Analysis: Detection, Processing, Classification and Applications. Biol. Proced. Online 8(1), 11–35 (2006)

[2] Kamen, G.: Frequently Asked Questions About Electromyography (EMG). In: Non-invasive Investigation of the Neuromuscular System, An Official Interest Group of ACSM. University of Massachusetts at Amherst, 25 de October del (2006), http://vuiis.vanderbilt.edu/~nins/EMG_FAQ.htm
[3] Betancourt, O.A., Suárez, E.G., Franco, J.F.: Movement Recognition from EMG Signals. Scientia et Técnica,Universidad Tecnológica de Pereira 10(26), 53–58 (2004)
[4] Berne, R.M., Levy, M.N.: Physiology: Section 1: Cellular Physiology. In: Ionic Equilibrium and Membrane Rest Potentia, Medica Panamericana, Buenos Aires Argentina, ch. 2, pp. 32–41 (1987)
[5] Merletti, R.: Standards for Reporting EMG Data, International Society of Electrophysiology and Kinesiology. Journal of Electromyography and Kinesiology, Politecnico di Torino, Italy (1999)
[6] Abbink, J.H., van der Bilt, A., van der Glas, H.W.: Detection of Onset and Termination of Muscle Activity in Surface Electromyograms. Journal of Oral Rehabilitation 25(5), 365–369 (1998)
[7] Merlo, A., Farina, D., Merletti, R.: A fast and reliable technique for muscle activity detection from surface EMG signals. IEEE Transactions on Biomedical Engineering 50(3), 316–323 (2003)
[8] Gazzoni, M., Farina, D., Merletti, R.: A New Method for the Extraction and Classification of Single Motor Unit Action Potentials From Surface EMG Signals. Journal of Neuroscience Methods 136(2), 165–177 (2004)
[9] Kiguchi, K., Tanaka, T., Fukuda, T.: Neuro-Fuzzy Control of a Robotic Exoskeleton With EMG Signals. IEEE Trans. Fuzzy Syst. 12(4), 481–490 (2004)
[10] Strang, G., Nguyen, T.: Wavelets and Filter Bank. Wellesley – Cambridge Press, USA (1996)
[11] Schrader, C.B., Roberson, D.: Using Rank Order Filters to Decompose the Electromyogram. In: Electronic Proceedings of 15th International Symposium on the Mathematical Theory of Networks and Systems, University of Notre Dame, South Bend, Indiana, USA, August 12-16 (2002)
[12] Liand, X., Aruin, A.S.: Muscle activity onset time detection using Teager-Kaiser energy operator. In: Proceedings of the 2005 IEEE Engineering in Medicine and Biology, 27th Annual Conference, September 2005, pp. 7549–7552 (2005)
[13] Linand, W., Chitrapu, P.: Time-Frequency Distributions Based on Teager-Kaiser Energy Function. In: Proceedings of 1996 IEEE International Conference on Acoustics, Speech, and Signal Processing, ICASSP 1996, May 1996, vol. 3, pp. 1818–1821 (1996)
[14] Vakman, D.: On the Analytic Signal, the Teager-Kaiser Energy Algorithm, and Other Methods for Defining Amplitude and Frequency. IEEE Transactions on Signal Processing 44(4), 791–797 (1996)
[15] Bezdek, J., Pal, K.: Fuzzy Models for Pattern Recognition. In: Methods that Search for Structures in Data. IEEE Press, Los Alamitos (1991)
[16] Pappas, T.: An adaptive Clustering Algorithm for Image Segmentation. IEEE Trans. Signal Processing 40, 901–914 (1992)

Part IV
Optimization and Robotics

Modular Neural Networks Architecture Optimization with a New Evolutionary Method Using a Fuzzy Combination Particle Swarm Optimization and Genetic Algorithms

Fevrier Valdez[1], Patricia Melin[1], and Guillermo Licea[2]

[1] Tijuana Institute of Technology, Tijuana BC. México
[2] Universidad Autónoma de Baja California
epmelin@hafsamx.org

Abstract. We describe in this paper a new hybrid approach for optimization combining Particle Swarm Optimization (PSO) and Genetic Algorithms (GAs) using Fuzzy Logic to integrate the results. The new evolutionary method combines the advantages of PSO and GA to give us an improved FPSO+FGA hybrid method. Fuzzy Logic is used to combine the results of the PSO and GA in the best way possible. Also fuzzy logic is used to adjust parameters in the FPSO and FGA. The new hybrid FPSO+FGA approach is compared with the PSO and GA methods for the optimization of modular neural networks. The new hybrid FPSO+FGA method is shown to be superior with respect to both the individual evolutionary methods.

1 Introduction

We describe in this paper a new evolutionary method combining PSO and GA, to give us an improved FPSO+FGA hybrid method. We apply the hybrid method to optimize the architectures of Modular Neural Networks (MNNs) for pattern recognition. We also apply the hybrid method to mathematical function optimization to validate the new approach. In this case, we are using the Rastrigin's function, Rosenbrock's function, Ackley's function, Sphere's function Griewank's function, Michalewics's function and Zakharov's function [18][16][17].

The paper is organized as follows: in part 2 a description about the genetic algorithms for optimization problems is given, in part 3 the Particle Swarm Optimization is presented, the FPSO+FGA method is presented in part 4 and part 5, the simulation results for modular neural network optimization are presented in part 6, finally we can see the conclusions reached after the study of the proposed evolutionary computing method.

P. Melin et al. (Eds.): Bio-inspired Hybrid Intelligent Systems, SCI 256, pp. 199–213.
springerlink.com

2 Genetic Algorithm for Optimization

John Holland, from the University of Michigan initiated his work on genetic algorithms at the beginning of the 1960s. His first achievement was the publication of *Adaptation in Natural and Artificial System* [11] in 1975.

He had two goals in mind: to improve the understanding of natural adaptation process, and to design artificial systems having properties similar to natural systems [10].

The basic idea is as follows: the genetic pool of a given population potentially contains the solution, or a better solution, to a given adaptive problem. This solution is not "active" because the genetic combination on which it relies is split between several subjects. Only the association of different genomes can lead to the solution.

Holland's method is especially effective because it not only considers the role of mutation, but it also uses genetic recombination, (crossover) [8]. The crossover of partial solutions greatly improves the capability of the algorithm to approach, and eventually find, the optimal solution.

The essence of the GA in both theoretical and practical domains has been well demonstrated [15]. The concept of applying a GA to solve engineering problems is feasible and sound. However, despite the distinct advantages of a GA for solving complicated, constrained and multi-objective functions where other techniques may have failed, the full power of the GA in application is yet to be exploited [3] [5].

In figure 1 we show the reproduction cycle of the Genetic Algorithm.

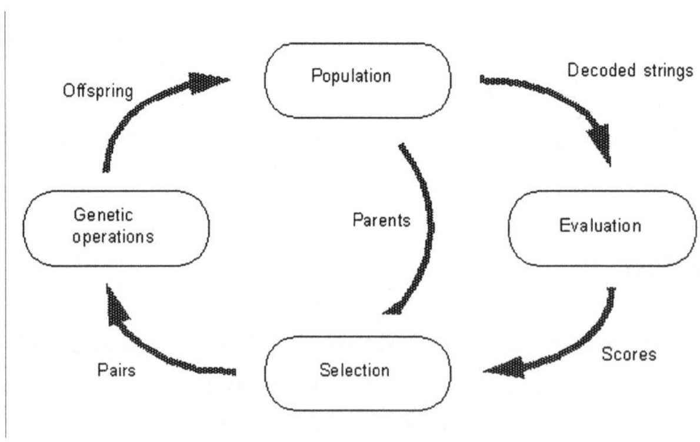

Fig. 1. The Reproduction cycle

The Simple Genetic Algorithm can be expressed in pseudo code with the following cycle:

1. Generate the initial population of individuals aleatorily P(0).
2. While (number _ generations <= maximum _ numbers _ generations)
  Do:
   {
    Evaluation;

Selection;
Reproduction;
Generation ++;
}
3. Show results
4. End

3 Particle Swarm Optimization

Particle swarm optimization (PSO) is a population based stochastic optimization technique developed by Eberhart and Kennedy in 1995, inspired by social behavior of bird flocking or fish schooling [14].

PSO shares many similarities with evolutionary computation techniques such as Genetic Algorithms (GA) [9]. The system is initialized with a population of random solutions and searches for optima by updating generations. However, unlike the GA, the PSO has no evolution operators such as crossover and mutation. In the PSO, the potential solutions, called particles, fly through the problem space by following the current optimum particles [1].

Each particle keeps track of its coordinates in the problem space, which are associated with the best solution (fitness) it has achieved so far (The fitness value is also stored). This value is called *pbest*. Another "best" value that is tracked by the particle swarm optimizer is the best value, obtained so far by any particle in the neighbors of the particle. This location is called *lbest*. When a particle takes all the population as its topological neighbors, the best value is a global best and is called *gbest* [12].

The particle swarm optimization concept consists of, at each time step, changing the velocity of (accelerating) each particle toward its p*best* and *lbest* locations (local version of PSO). Acceleration is weighted by a random term, with separate random numbers being generated for acceleration toward *pbest* and *lbest* locations [12].

In the past several years, PSO has been successfully applied in many research and application areas. It is demonstrated that PSO gets better results in a faster, cheaper way compared with other methods [2] [13].

Another reason that PSO is attractive is that there are few parameters to adjust. One version, with slight variations, works well in a wide variety of applications. Particle swarm optimization has been used for approaches that can be used across a wide range of applications, as well as for specific applications focused on a specific requirement [2][20].

The pseudo code of the PSO is as follows

For each particle
Initialize particle
End
Do
For each particle
Calculate fitness value
If the fitness value is better than the best fitness value (pBest) in history
set current value as the new pBest

End
Choose the particle with the best fitness value of all the particles as the gBest
For each particle
Calculate particle velocity
Update particle position
End
While maximum iterations or minimum error criteria is not attained

4 FPSO+FGA Method

This method combines the characteristics of PSO and GA using several fuzzy systems for integration of results and parameter adaptation. In this section, the proposed FPSO+FGA method is presented.

The general idea of the proposed FPSO+FGA method can be seen in figure 2. The method can be described as follows:

1. A mathematical function to be optimized is received.
2. The role of both FPSO and FGA is evaluated.
3. A main fuzzy system is responsible for receiving values resulting from step 2.

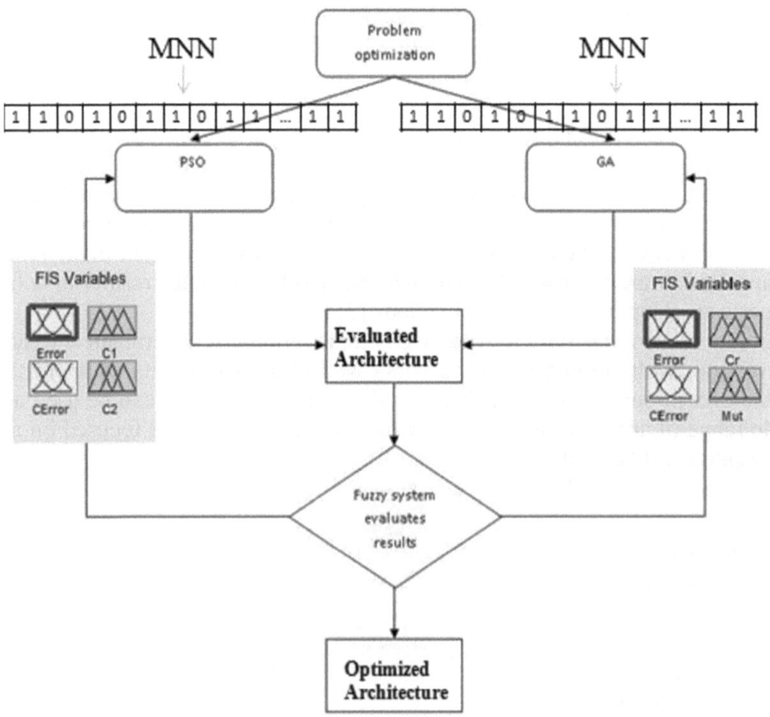

Fig. 2. The FPSO+FGA scheme

4. The main fuzzy system decides which method to use (FPSO or FGA)
5. Another fuzzy system receives the values of Error and DError as inputs to evaluate if it is necessary to change the parameters in FPSO or FGA.
6. There are 3 fuzzy systems. One is for decision making (is called 'fuzzy-main'), the second one is to change the parameters of the GA (is called 'fuzzyga'), in this case change the value of the crossover and mutation rate and the third fuzzy system is used to change the parameters of the PSO (is called 'fuzzypso') in this case change the values of the cognitive acceleration 'c1', and social acceleration 'c2'. The main fuzzy system (called 'fuzzymain') decides in the final step the optimum value for the function introduced in step 1.
7. Repeat the above steps until the termination criterion of the algorithm is met.

5 Full Model FPSO+FGA

The basic idea of the FPSO+FGA scheme is to combine the advantage of the individual methods using a fuzzy system for decision making and the others two fuzzy systems to improve the parameters of the FGA and FPSO when is necessary.

As can be seen in the proposed hybrid FPSO+FGA method, it is the internal fuzzy system structure, which has the primary function of receiving as inputs (Error and DError) the results of the FGA and FPSO outputs. The fuzzy system is responsible for integrating and decides which are the best results being generated at run time of the FPSO+FGA. It is also responsible for selecting and sending the problem to the "fuzzypso" fuzzy system when the FPSO is activated or to the "fuzzyga" fuzzy system when FGA is activated. Also activating or temporarily stopping depending on the results being generated. Figure 4 shows the membership functions of the main fuzzy system that is implemented in this method. The fuzzy system is of Mamdani type because it is more common in this type of fuzzy control and the defuzzification method is the centroid. In this case, we are using this type of defuzzification because in other papers we have achieved good results [4]. The membership functions are of triangular form in the inputs and outputs as is shown in figure 4. Also, the membership functions were chosen of triangular form based on past experiences in this type of fuzzy control. The fuzzy system consists of 9 rules. For example, one rule is if error is P and DError is P then best value is P (view figure 5). Figure 6 shows the fuzzy system rule viewer. Figure 7 shows the surface corresponding to this fuzzy system. The other two fuzzy systems are similar to the main fuzzy system.

5.1 FPSO (Fuzzy Particle Swarm Optimization)

This section presents a detailed description of the FPSO model. The classical representation scheme for GAs is binary vectors of fixed length. In the case of an n_x – dimensional search space, each individual consists of n_x variables with each variable encoded as a binary string.

The swarm is typically modeled by particles in multidimensional space that have a position and a velocity. These particles fly through hyperspace (i.e., R^n) and have two essential reasoning capabilities: their memory of their own best position and knowledge of the global or their neighborhood's best. In figure 3 we can see a simulation of Ackley's function. In a minimization optimization problem, "best" simply meaning the position with the smallest objective value. Members of a swarm communicate good positions to each other and adjust their own position and velocity based on these good positions. So a particle has the following information to make a suitable change in its position and velocity:

- A global best that is known to all and immediately updated when a new best position is found by any particle in the swarm.
- The neighborhood best that the particle obtains by communicating with a subset of the swarm.
- The local best, which is the best solution that the particle has seen.

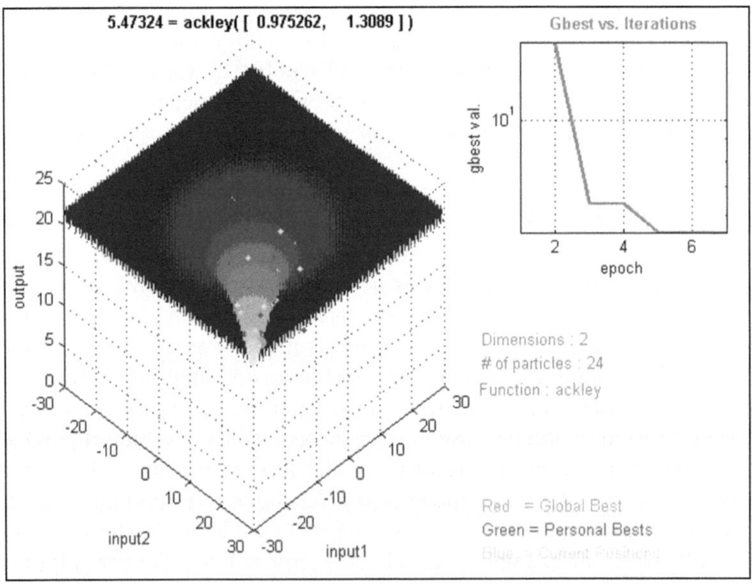

Fig. 3. Simulation of ackley's function with FPSO

In this case, the social information is the best position found by the swarm, referred as $\hat{y}(t)$. For gbest FPSO, the velocity of particle i is calculated as

$$v_{ij}(t+1) = vi_j(t) + c_1 r_{1j}(t)[y_{ij}(t) - x_{ij}(t)] + c_2 r_{2j}(t)[\hat{y}_j(t) - x_{ij}(t)] \quad (1)$$

Where $vi_j(t)$ is the velocity of particle i in dimension j = 1,…,n_x at time step t, $xi_j(t)$ is the position of particle i in dimension j at time step t, C_1 and C_2 represents the cognitive and social acceleration. In this case, these values are fuzzy because they are changing dynamically when the FPSO is running, and r_{1j}(t), r_{2j}~U(0,1) are random values in the range [0,1].

5.2 FGA (Fuzzy Genetic Algorithm)

This section presents a detailed description of the FGA. Several crossover operators have been developed for GAs, depending on the format in which individuals are represented. For binary representations, uniform crossover, one point crossover and two points cross over are the most popular. In this case we are using two points crossover with fuzzy crossover rate because we are adding a fuzzy system called 'fuzzyga' that is able of change the crossover and mutation rate.

5.3 Definition of the Fuzzy Systems Used in FPSO+FGA

'fuzzypso': In this case we are using a fuzzy system called 'fuzzypso', and the structure of this fuzzy system is as follow:

Number of Inputs: 2
Number of Outputs: 2
Number of membership functions: 3
Type of the membership functions: Triangular
Number of rules: 9

Defuzzification: Centroid
The main function of the fuzzy system called 'fuzzypso' is to adjust the parameters of the PSO. In this case, we are adjusting the following parameters: 'c1' and 'c2' ; where:
'c1' = Cognitive Acceleration
'c2' = Social Acceleration

We are changing these parameters to test the proposed method. In this case, with 'fuzzypso' is possible to adjust in real time the 2 parameters that belong to the PSO.

'fuzzyga': In this case we are using a fuzzy system called **'fuzzyga'**, the structure of this fuzzy system is as follows:

Number of Inputs: 2
Number of Outputs: 2
Number of membership functions: 3

Type of membership functions: Triangular
Number of rules: 9

Defuzzification: Centroid
The main function of the fuzzy system called 'fuzzypso' is to adjust the parameters of the GA. In this case, we are adjusting the following parameters: 'mu', 'cr'; where:

'mu' = mutation
'cr' = crossover

'fuzzymain': In this case, we are using a fuzzy system called **'fuzzymain'**. The structure of this fuzzy system is as follows:

Number of Inputs: 2
Number of Outputs: 1
Number of membership functions: 3
Type of membership functions: Triangular
Number of rules: 9

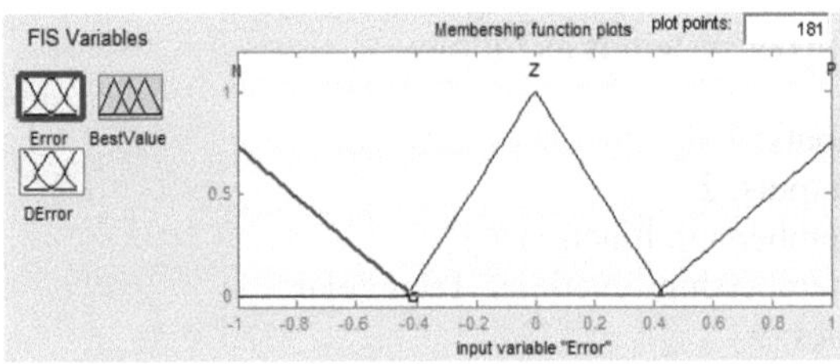

Fig. 4. Fuzzy system membership functions

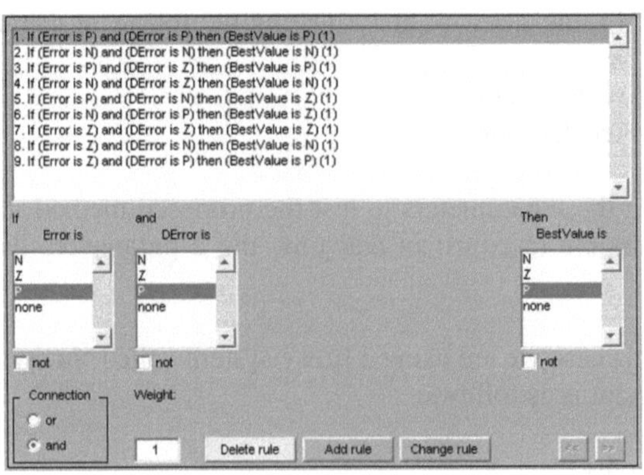

Fig. 5. Fuzzy system rules

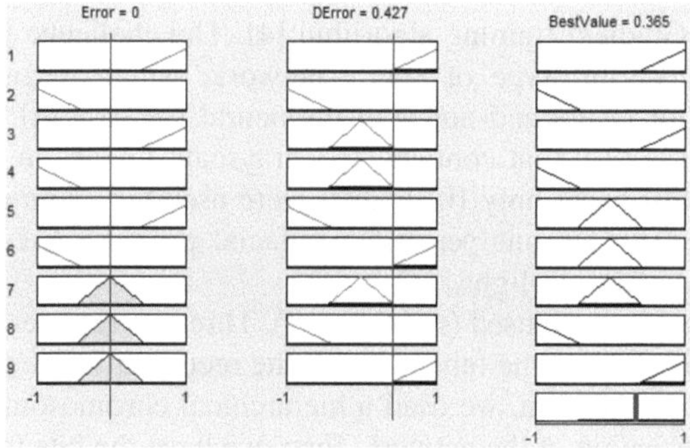

Fig. 6. Fuzzy system rules viewer

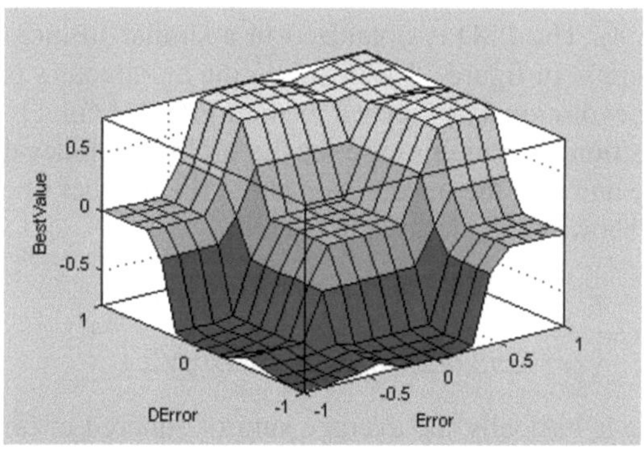

Fig. 7. Surface of fuzzy system

Defuzzification: Centroid

The main function of the fuzzy system, called 'fuzzymain' is to decide on the best way for solving the problem, in other words if it is more reliable to use the FPSO or FGA. This fuzzy system is able to receive two inputs, called error and derror, it is to evaluate the results that are generated by FPSO and FGA in the last step of the algorithm.

6 Simulation Results for Modular Neural Network for Optimization

Several tests of the FPSO+FGA method for MNN optimization were made in the Matlab programming language.

All the implementations were developed using a computer with quad core2 processor of 64 bits that works to a frequency of clock of 2.5 GHz, 4 GB of RAM Memory and Windows Vista operating system.

We describe below simulation results of our approach for face recognition with modular neural networks (MNNs). We used two-layer feed-forward MNNs with the Conjugated-Gradient training algorithm [4]. The challenge is to find the optimal architecture of this type of neural network, which means finding out the optimal number of layers and nodes of the neural network [6]. We are using the Yale face database [20] that contains 165 grayscale images in GIF format of 15 individuals, for this paper only 10 subjects were used for training the MNN. There are 5 images per subject, one per different facial expression: center-light, happy, left-light, normal and right-light.

In total 50 images were used (see figure 9). Three images per subject were used for training the MNN and the other two for the recognition. Regarding the genetic algorithm for NN evolution, we used a hierarchical chromosome for representing the relevant information of the network. First, we have the bits for representing the number of layers of the MNN, in this case, the initial topology was of 3 modules with 2 layers per module with 500 neurons in the first layer, 300 neurons in the second layer in each module. Therefore we used a representation the 2415 bits in total (view figure 8). The PSO is organized in a similar fashion, but there is less number of parameters. In figure 10 we can see the architecture of a MNN that we are using with the evolutionary proposed method FPSO+FGA.

The fitness function used in this case for the MNN combines the information of the error objective and also the information about the number of nodes as a second objective. This is shown in the following equation.

$$f(z) = \left(\frac{1}{\alpha * Ranking(ObjV1) + \beta * ObjV2} \right) * 10 \tag{2}$$

The first objective is basically the average sum of squared of errors as calculated by the predicted outputs of the MNN compared with real values of the function. This is given by the following equation.

$$f_1 = \frac{1}{N} \sum_{i=1}^{N} (Y_i - y_i)^2 \tag{3}$$

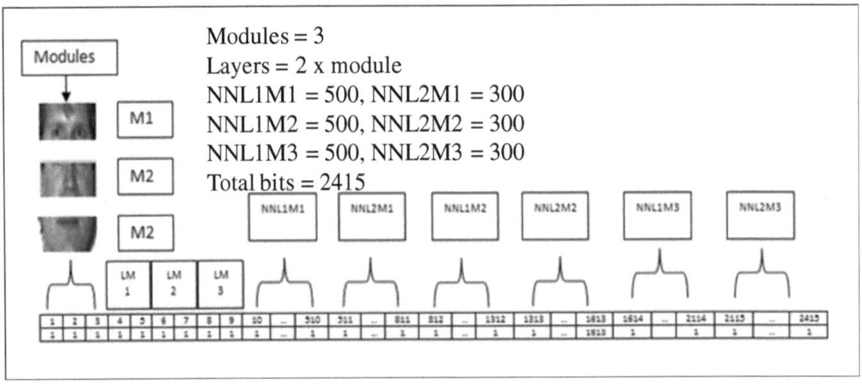

Fig. 8. Binary representation for FPSO+FGA (No optimized)

Fig. 9. Images of the Yale face database

The second objective is the complexity of the neural network, which is measured by the total number of nodes in the architecture.

The final topology of the neural network for the problem of face recognition is obtained by the hybrid evolutionary method FPSO+FGA. The comparison of the final objective values (errors) will be shown in the following section. In the final architecture, the result of the MNN evolution is a particular architecture with different number of nodes by layers. Several tests were made; we obtained different optimized architectures for this Modular Neural Network; the best architecture obtained was the following:

Layers = 2 x module
NNL1M1 = 90, NNL2M1 = 50
NNL1M2 = 100, NNL2M2 = 150
NNL1M3 = 70, NNL2M3 = 90
Total bits = 565
Where:
NNL1M1= Number of neurons of the layer 1 in module 1.
NNL1M1= Number of neurons of the layer 1 in module 1.
NNL2M1= Number of neurons of the layer 2 in module 1.
NNL1M2= Number of neurons of the layer 1 in module 2.
NNL2M2= Number of neurons of the layer 2 in module 2.
NNL1M3= Number of neurons of the layer 1 in module 3.
NNL2M3= Number of neurons of the layer 2 in module 3.

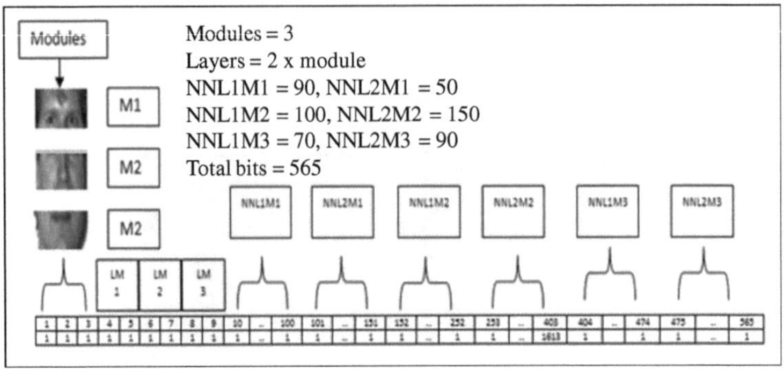

Fig. 10. Binary representation optimized for FPSO+FGA.

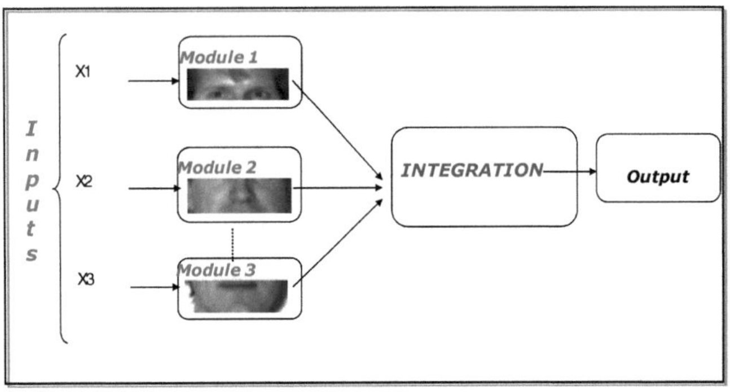

Fig. 11. Architecture of the Modular Neural Network.

We can see in the figure 10 the binary representation for this optimized architecture. With this final topology the Neural Network was trained and the ten images were recognized. It can be seen in table I the different architectures obtained with this method. The proposed method optimizes the initial architecture proposed for the problem of face recognition.

The Parameters of Table 1 and 2, are as follows:
Ima = Number of images. Mod = Number of modules. NNL = Number of neurons Layer 1, 2 or 3. GE = Goal Error. RE = Reached Error. IDENT = Number of images recognized. Train M = Training Method for the MNN. FGA = Parameters of Fuzzy Genetic Algorithm. FPSO = Parameters of Fuzzy Genetic Algorithm. VAR = Number of variables for the mathematical function.

Also this FPSO+FGA have been applied, for optimization of complex mathematical functions to validate our approach. The table 2, shows the simulation results with 7 mathematical functions. It can be seen in table 2 that this method is good alternative to solve this type of problems. The mathematical functions are evaluated with 2, 4, 8 and 16 variables.

Table 1. Simulation Results for the MNN

Mod	LMod	NN L1M1	NN L2M1	NN L1M2	NN L2M2	NN L1M3	NN L2M3	GE	RE	IDENT
3	2	20	60	80	50	60	120	0.01	0.03	8
3	2	90	50	100	150	70	90	0.01	0.005	10
3	2	70	40	80	40	90	30	0.01	0.02	8
3	2	150	135	200	90	84	40	0.01	0.003	9
3	2	100	120	100	145	100	70	0.01	0.001	10

Table 2. Simulation results for Mathematical Functions

Math function	**Var = 2**		**Var = 4**		**Var = 8**		VAR = 16	
Math	BEST	MEAN	BEST	MEAN	BEST	MEAN	BEST	MEAN
Rastrigin	1.45E-06	3.05E-04	0.00034	0.0755	0.01318	0.9311	0.548397150613653	5.0946
Rosen-brock	1.17E-02	1.17E-02	0.0285	0.5991	0.15800	3.8925	0.25555	4.33334
Ackley	8.42E-04	4.98E-03	8.42E-01	4.98E-02	0.7	1.56	2.35	2.63
Sphere	5.75E-11	1.05E-10	1.946e-05	4.5109e-004	0.00059	0.0057	0.00248	0.0211
Griewank	7.88E-11	1.07E-07	7.18130761856450e-06	1.1182e-004	0.000162543802633697	9.2993e-004	0.000408945493635127	0.0043
Michale-wics	-1.8010	-1.8201	-1.80129624818704	-1.8002	-1.80130169498191	-1.8005	-1.80130138163081	-1.8004
Zakharov	6.00E-07	0.00168	3.23737992547765e-07	8.4129e-005	1.33084082865516e-07	6.4901e-005	8.63410514832145e-07	7.0065e-005

7 Conclusions

The analysis of the simulation results of the evolutionary method considered in this paper, FPSO+FGA lead us to the conclusion that for the optimization of Modular Neural Networks with this method is a good alternative because it is easier to optimize the architecture of Modular Neural Network than to try it with PSO or GA separately. This is, because the combination PSO and GA with fuzzy rules gives a new hybrid method FPSO+FGA. It can be seen in table I that the second and fourth architectures obtained after applying FPSO+FGA recognize the ten images, and as a consequence we are demonstrating that it is reliable for this type of applications. Recently we are working with more images to test the effectiveness of this approach. Also this method has been tested with other mathematical functions.

Acknowledgment

We would like to express our gratitude to CONACYT, Universidad Autónoma de Baja California and Tijuana Institute of Technology for the facilities and resources granted for the development of this research.

References

[1] Angeline, P.J.: Using Selection to Improve Particle Swarm Optimization. In: Proceedings 1998 IEEE World Congress on Computational Intelligence, Anchorage, Alaska, pp. 84–89. IEEE, Los Alamitos (1998)

[2] Angeline, P.J.: Evolutionary Optimization Versus Particle Swarm Optimization: Philosophy and Performance Differences. In: Porto, V.W., Waagen, D. (eds.) EP 1998. LNCS, vol. 1447, pp. 601–610. Springer, Heidelberg (1998)

[3] Back, T., Fogel, D.B., Michalewicz, Z. (eds.): Handbook of Evolutionary Computation. Oxford University Press, Oxford (1997)

[4] Castillo, O., Huesca, G., Valdez, F.: In: Proceedings of the International Conference on Artificial Intelligence, IC-AI 2004, June 21-24, vol. 1, pp. 98–104 (2004)

[5] Castillo, O., Valdez, F., Melin, P.: Hierarchical Genetic Algorithms for topology optimization in fuzzy control systems. International Journal of General Systems 36(5), 575–591 (2007)

[6] Castillo, O., Melin, P.: Hybrid intelligent systems for time series prediction using neural networks, fuzzy logic, and fractal theory. IEEE Transactions on Neural Networks 13(6), 1395–1408 (2002)

[7] Eberhart, R.C., Kennedy, J.: A new optimizer using particle swarm theory. In: Proceedings of the Sixth International Symposium on Micromachine and Human Science, Nagoya, Japan, pp. 39–43 (1995); Lu, J.-G.: Title of paper with only the first word capitalized. J. Name Stand. Abbrev. (in press)

[8] Emmeche, C.: Garden in the Machine. In: The Emerging Science of Artificial Life, p. 114. Princeton University Press, Princeton (1994)

[9] Fogel, D.B.: An introduction to simulated evolutionary optimization. IEEE transactions on neural networks 5(1), 3–14 (1994)

[10] Goldberg, D.: Genetic Algorithms. Addison-Wesley, Reading (1988)

[11] Holland, J.H.: Adaptation in natural and artificial system. The University of Michigan Press, Ann Arbor (1975)

[12] Kennedy, J., Mendes, R.: Population structure and particle swarm performance. In: Proceeding of IEEE conference on Evolutionary Computation, pp. 1671–1676 (2002)

[13] Kennedy, J., Mendes, R.: The particle swarm-explosion, stability, and convergence in a multidimensional complex space. IEEE Transactions on Evolutionary Computation 6(1), 58–73 (2002)
[14] Kennedy, J., Eberhart, R.C.: Particle swarm optimization. In: Proceedings of IEEE International Conference on Neural Networks, Piscataway, NJ, pp. 1942–1948 (1995)
[15] Man, K.F., Tang, K.S., Kwong, S.: Genetic Algorithms: Concepts and Designs. Springer, Heidelberg (1999)
[16] Montiel, O., Castillo, O., Melin, P., Rodriguez, A., Sepulveda, R.: Human evolutionary model: A new approach to optimization. Inf. Sci. 177(10), 2075–2098 (2007)
[17] Valdez, F., Melin, P., Castillo, O.: Evolutionary Computing for the Optimization of Mathematical functions. In: Analysis and Design of Intelligent Systems Using Soft Computing Techniques. Advances in Soft Computing, vol. 41 (June 2007)
[18] Valdez, F., Melin, P.: Parallel Evolutionary Computing using a cluster for Mathematical Function Optimization. In: NAFIPS, San Diego CA USA, June 2007, pp. 598–602 (2007)
[19] Veeramachaneni, K., Osadciw, L., Yan, W.: Improving Classifier Fusion Using Particle Swarm Optimization. In: IEEE Fusion Conference, Italy (July 2006)
[20] Wei, W., Jiatao, S., Zhongzxiu, Y., Zheru., C.: Wavelet-based Illumination Compensation for Face Recognition using Eifenface Method Intelligent Control and Automation. In: WCICA 2006, 21-23 June 2006, vol. 2, pp. 10356–10360 (2006)

A Multi-agent Architecture for Controlling Autonomous Mobile Robots Using Fuzzy Logic and Obstacle Avoidance with Computer Vision

Cinthya Solano-Aragón[1] and Arnulfo Alanis[2]

[1] Student of Master of Science in Computer Science,
Instituto Tecnologico de Tijuana, Calzada Tecnologico S / N, Tijuana, Mexico
cinthyasolanoa@gmail.com
[2] Division of Graduate Studies and Research, Instituto Tecnologico de Tijuana,
Calzada Tecnologico, S/N, Tijuana, Mexico
alanis@tectijuana.edu.mx

Abstract. This paper describes the development of a Multi-Agent System (MAS), which is supported with fuzzy logic (to control the robots movements in a reactive path) and computer vision, which controls an autonomous mobile robot to exit a maze. The research consists of two stages. In the first stage the problem is to be able to make the robot exit a maze, the mobile robot is positioned at the entrance (point A) and should reach an output (B), it should be noted that we are working with a NXT robot to Lego MINDSTORMS ®. In its second phase the problem is to make the robot search for a recognized object, for this purpose a camera is used to capture images, which will be processed with vision techniques for identification, and after that, the SMA takes the decision to evade or take the object as appropriate.

1 Introduction

Making smarter robots is a problem that has captivated the scientific community for several years now, this being a big challenge to overcome even for man. The move from one place to another from an initial to a final point with only the information of what we want to reach at the end is a complicated task. The problem has been solved with soft computing methods, such as fuzzy logic, neural networks, genetic algorithms, hybrid systems, among others. Navigation of the robot has been achieved taking a completely controlled environment where we have total knowledge of the world [7, 10, 15, 16].

Human beings have the ability to react to unexpected situations and the ability to use their reasoning to react to situations that avise, an example is a situation of traffic management, to which we think and what is the best route and with this the rest of the trip is done reacting to what was observed. This could be, at a traffic light, when it changes from green to yellow, the reaction will be learning, acceleration or deceleration, a situation that when you start, it has to be learned.

P. Melin et al. (Eds.): Bio-inspired Hybrid Intelligent Systems, SCI 256, pp. 215–246.
springerlink.com

Knowing that we have this ability to react, we will use fuzzy logic to transfer the knowledge we use to carry out this process by establishing computer-type "if-then" fuzzy rules and exploiting the advantages offered to us by fuzzy logic, which is the use of linguistic variables.

A man driving a car as in many other activities derives its information by looking at distances, clearances and other identifying information that we collect through the light. In this research we use computer vision techniques to draw from that vital information for navigation control, supported with special sensors. An example is the ultrasonic sensors that are used to measure distance and the light sensors that we can use to measure the change in light intensity, all of these help to extract environmental information necessary for decision making.

2 Agents

Let's first deal with the notion of intelligent agents. These are generally defined as "software entities", which assist their users and act on their behalf. Agents make your life easier, save you time, and simplify the growing complexity of the world, acting like a personal secretary, assistant, or personal advisor, who learns what you like and can anticipate what you want or need. The principle of such intelligence is practically the same of human intelligence. Through a relation of collaboration-interaction with its user, the agent is able to learn from himself, from the external world and even from other agents, and consequently act autonomously from the user, adapt itself to the multiplicity of experiences and change its behavior according to them. The possibilities offered for humans, in a world whose complexity is growing exponentially, are enormous [42][43][44][45].

We need to be careful to distinguish between rationality and omniscience. An omniscient agent knows the actual outcome of its actions, and can act accordingly; but omniscience is impossible in reality. Consider the following example: I am walking along the Champs Elys´ees one day and I see an old friend across the street. There is no traffic nearby and I'm not otherwise engaged, so, being rational, I start to cross the street. Meanwhile, at 33,000 feet, a cargo door falls off a passing airliner, and before I make it to the other side of the street I am flattened. Was I irrational to cross the street? It is unlikely that my obituary would read "Idiot attempts to cross street."Rather, this point out that rationality is concerned with expected success given what has been perceived. Crossing the street was rational because most of the time the crossing would be successful, and there was no way I could have foreseen the falling door. Note that another agent that was equipped with radar for detecting falling doors or a steel cage strong enough to repel them would be more successful, but it would not be any more rational [23].

3 FIPA

FIPA (The Foundation of Intelligence Physical Agents). FIPA specifications represent a collection of standards, which are intended to promote the interoperation of heterogeneous agents and the services that they can represent.

The life cycle [46] of specifications details what stages a specification can attain while it is part of the FIPA standards process. Each specification is assigned a specification identifier [47] as it enters the FIPA specification life cycle. The specifications themselves can be found in a Repository [11]. The Foundation of Intelligent Physical Agents (FIPA) is now an official IEEE Standards Committee. The UML modeling language is the most popular system used today and it is a graphic language to visualize, specify, build and document a system. Gaia is a methodology for application development in the paradigm of agents, one of the most curious aspects of Gaia, is the fact that the requirements specification is completely independent of the analysis and design.

4 Proposed Method

To realize this work, the creation of a SMA is needed, which is composed of three agents, an agent for the management of sensors, an agent for handling the drive and a last agent to act as coordinator of the multi-agent system.

Summarizing the SMA is composed of the following agents:

Node Agent (NA), a set of sensors.
Task Agent (TA), a set of servo motors.
Agent System (AS), coordinator of the system.

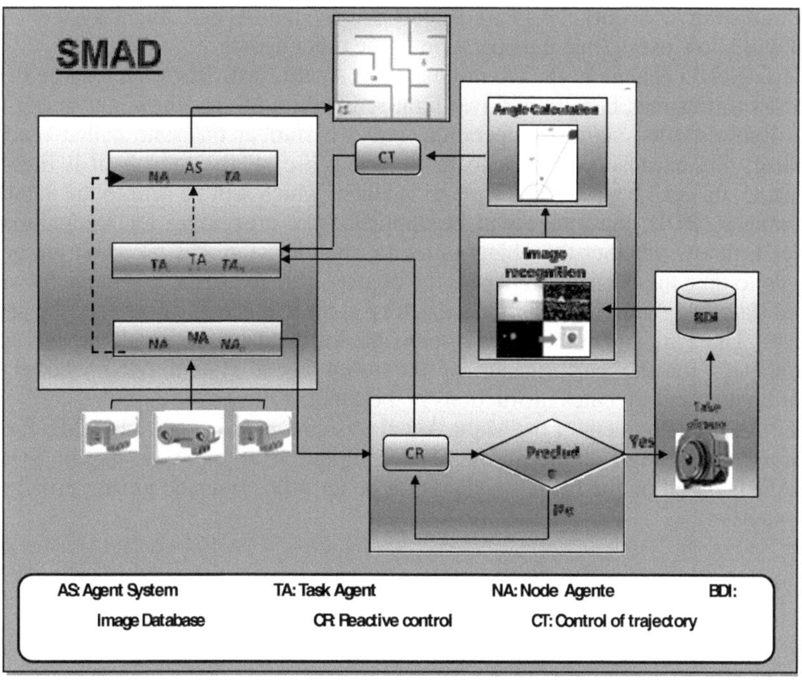

Fig. 1. Architecture of the complete system.

The Node Agent (NA) will be responsible for the sensors used by our robot that are:

- Ultrasonic Sensor
- Light Sensor left
- Light Sensor right
- Camera

These sensors are used to interact with the world to achieve our goal.

The Task Agent (TA) will be used to drive to move forward, rewind, or make some movement to escape.

The System Agent (SA) will be responsible for coordinating the other actors, the AN and AT, and will also have activities not only the coordination, which will be to recognize the object that is opposite, calculate the angle of the object, which will help us to know that will guide the robot. Depending on the recognition that was obtained, whether or not the object, which will tell us that there is no escape or the object based on this decision was to trigger the switch which will control the outcome to be used, reactive or path to better performance of the controllers.

In Fig. 1 we show graphically the architecture of the complete system.

We can describe the proposed method as follows:

In the operation of the model, we have 2 main blocks, which are responsible for knowledge and learning (paradigm of intelligent agents) and the vision and control (fuzzy logic).

These modulus are described below, the first module contains 3 agents, NA [Administrative Agent], TA [Task Agent] SA [System Agent] Agent System (AS), and will know every time the operation of the other agents.

To detect a change in the environment the Agent Node [that is in charge of the sensors (ultrasonic, camera and two light sensors)] with the ultrasonic sensor and two light sensors, starts its operation, and we start at the state called reactive control; this because you have to move and sense all the time until it finds an obstacle, this can happen once a photo is taken, which will be sent to the database of images (BDI), the image will be applied a pre-processing for recognition in order to know whether the object is found, this decision was taken on the angle (angle to take the decision to bypass the object), able to escape and move forward, trend data, they are caught in the trajectory control process, to observe the speed values. The Task Agent (which is responsible for the drive), once all the necessary parameters for the performance of the robot agent system (AS) who is the coordinator, and executes instructions in the robot.

For the development of the Multi-Agent System we used Gaia and UML for the analysis and design of the agents. The completion of the Multi-Agent System (MAS) is based on the FIPA standards for better utilization and greater possibility of extension.

5 Analysis and Design of the Multi-Agent System (MAS)

The first activity was to perform the analysis and design of the multi-agent system as mentioned above, and to develop the two tools we describe the following concepts.

5.1 UML

Use Case Diagrams

The Use Case diagrams show the granularity of the system into reusable pieces of functionality, interaction of players with the functionality of the system, visually organize user requirements and allow the contract to certify the functionality, and formalize the process map.

Fig. 2 presents the use case of the sense in which the distance is appreciated that other processes must be carried out before and after to complete. In Fig. 3 shows the use case of the motor A move, which is seen in all other processes that must be carried out before and after to complete.

Fig. 4 presents the case for use of the remote sensing process, which shows that other processes must be carried out before and after to complete. Fig. 5 shows the use case of the motor A move, which is seen in all other processes that must be carried out before and after to complete.

Sequence diagrams describe the interaction of objects that require the functionality of the different scenarios of a use case, objects are represented with their life cycle within a time series, and each possible scenario of a use case can be represented as a sequence diagram.

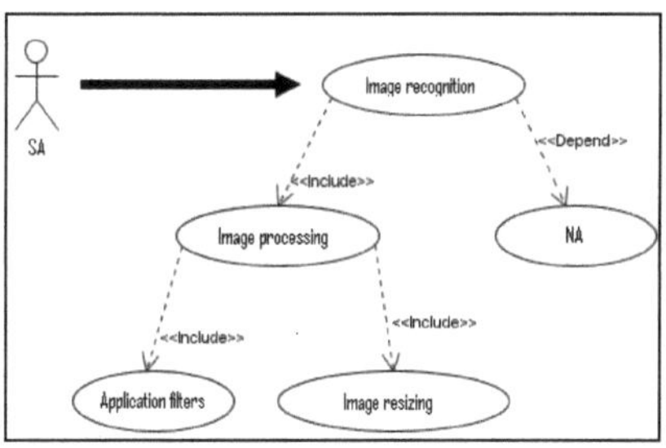

Fig. 2. Use case diagram of the process of image recognition.

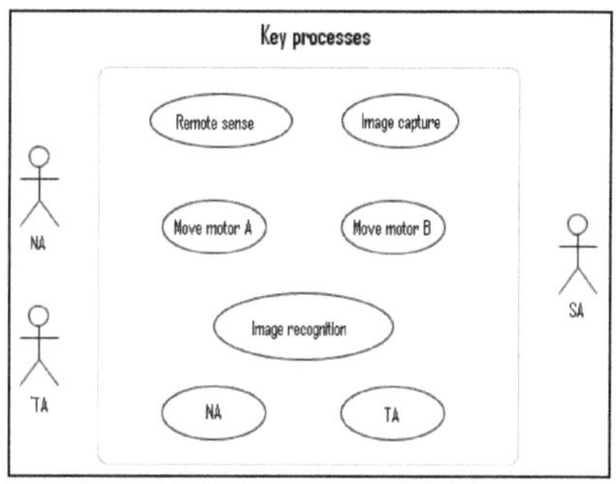

Fig. 3. Main processes of use cases.

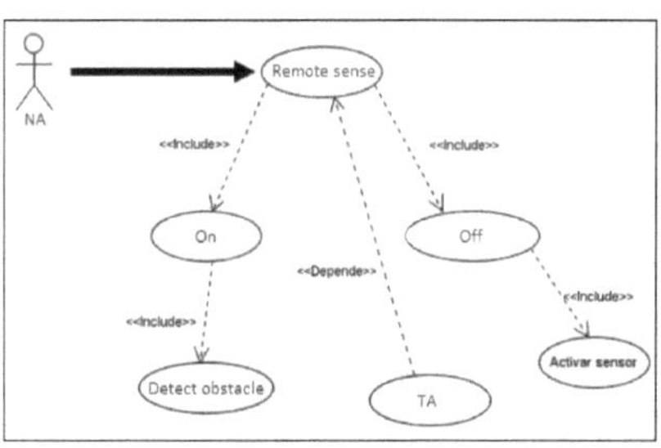

Fig. 4. Use case diagram of the remote sense.

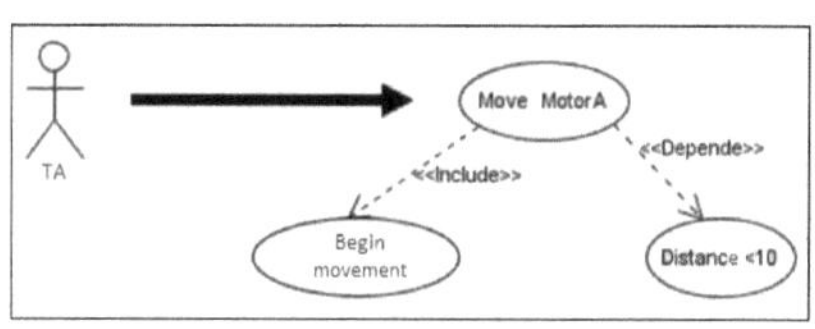

Fig. 5. Use case diagram of the process moving motor A.

Below in Fig. 6 we show the sequence diagram of the use case for moving motor A. In Fig. 7 presents the sequence diagram of use case to capture the image which is carried out by the agent node (AN).

In Fig. 8 we show the sequence diagram used for the case sense of distance which is carried out by the agent node (AN).

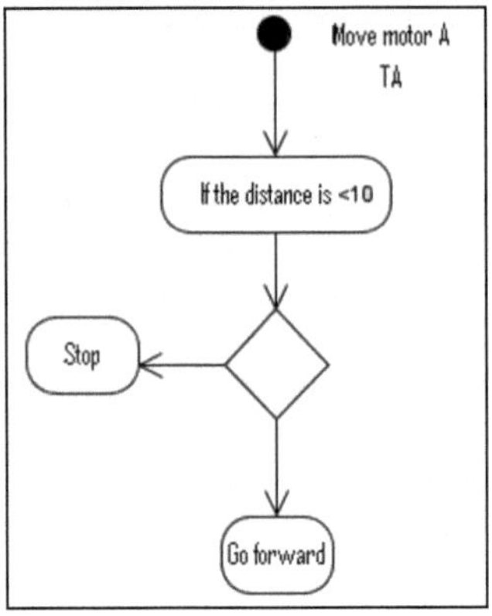

Fig. 6. Sequence diagram use case to move motor A.

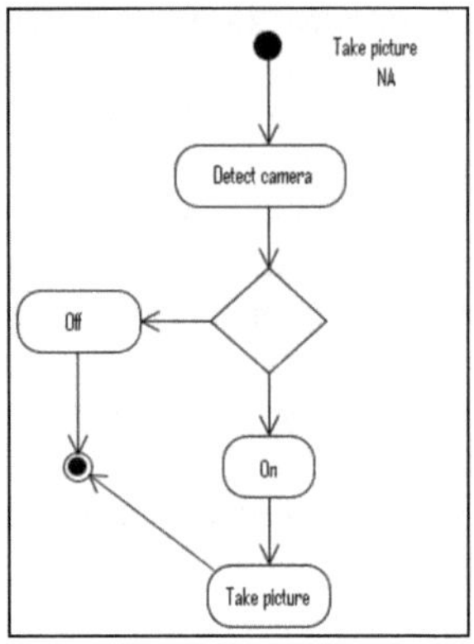

Fig. 7. Sequence diagram use case capture image.

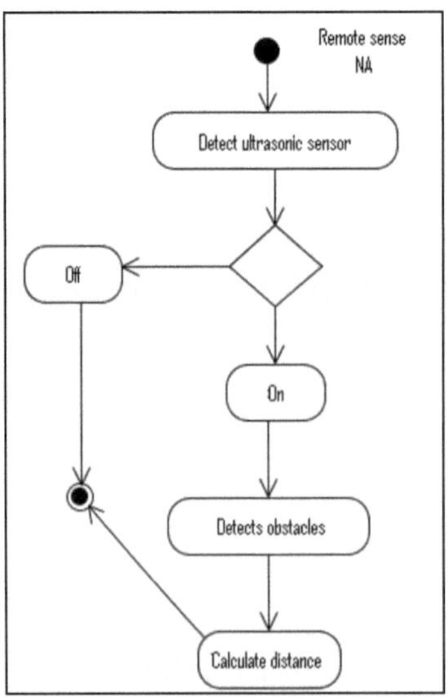

Fig. 8. Sequence diagram use case sense distance.

5.2 Gaia

Below we show the diagrams in which we can observe the responsibilities, permissions, activities and protocols to be followed by each agent in the Multi-Agent System (MAS).

Agent node

Responsibilities:
Sense:
Active
Inactive

Permissions:
Ultrasonic sensor: measures distances from 0 cm to 255 cm, with a delay of one millisecond signal.
Camera Sensor: take pictures in an estimated time of seconds, saves all images in a folder for processing. It has a degree of vision approximately of 50 degrees.
Light sensors: they measure the intensity with which an object reflects light. This makes a light emitting and measuring the portion of the return is received, the table 3.2 shows the ranges of values.

Activities:
The only activities are sensing and taking pictures of the world.

Protocols:
Ultrasound: This is responsible for sensing the distance to get to an obstacle. The initiative for this would be the agent system, which indicates the time of initiation.
Chamber is responsible for obtaining the images within the scene, and as for the ultrasonic sensor system depends for its initiation.
Light sensors: the role of these sensors is to measure the intensity with which an object reflects light (walls, diagrams). This makes a light emitting and measuring the portion of the return is received and handled the ranges are 0 to 1023 RAW.

Agent system

Responsibilities:
It is responsible for image processing and decision making, and sends a message to agent communication process completed?

Permissions:
This is the one that has full access, sensors, communication, engine and handling agents and NXT in general.

Activities:
Making decisions based on the behavior and implements what is needed to finish the job.

Protocols:

Full Access, sensors, communication, engine and handling agents and NXT in general, the entry for this information would be provided by the agents, getting a response from these so they generate an output and generate interaction among them.

Task agent

Responsibilities:

Motor movement

Permission:

Depends entirely on the agent system to perform its task.

Activities:

Sensors for the engine: they measure in degrees is given for each engine is 360 degrees for one revolution, a revolution for the tire of each motor, the motors can rotate independently, can also be used with constant speed motors.

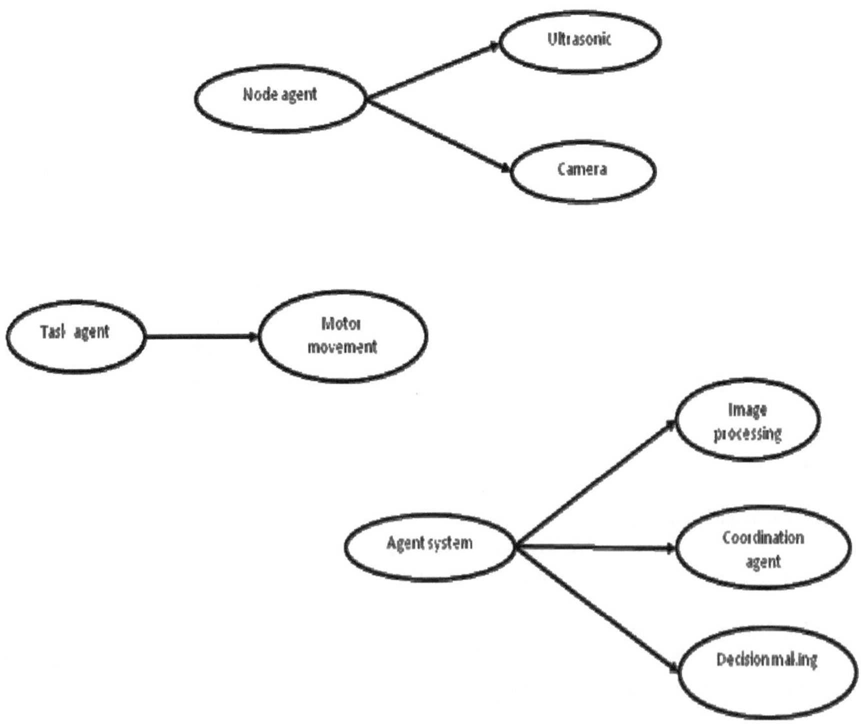

Fig. 9. Shows the diagrams of the Agent Node (AN), Task Agent (TA), and Agent System (AS) in Gaia.

Fig. 10. Diagram of the relationship model of SMA.

Protocols:

The system is the initiator, resulting in a movement or departure, which can be any direction, depending on the decision taken by the agent system.

Figure 9 shows the three models that form the multi-agent system (MAS) using Gaia.

Fig. 10 presents the relationship between the agents of the multi-agent system (MAS).

5.3 Knowledge Base

The knowledge base with which the MAS is working, is shown below.

Facts

AN_i.SU. STATE =(X), where X can have 2 states (ON, OFF)
AN_i.SU.OBJ=(Y) where Y can have 2 states (DETECTS, NO DETECTS)
AN_i.SC. STATE = (Z) where Z can have 2 states (ON, OFF)
AN_i.SC.TIMAGE
AN_i.SLI. STATE = (W) where W can have 2 states (ON, OFF)
AN_i.SLD. STATE = (V) where V can have 2 states (ON, OFF)
AT_j.M.GO(evaluationFIS[SU, SLI, SLD]) where SU,SLI,SLD FIS are entries for return rates of both servomotors (MA, MB)
AT_j.M.STOP
AT_j.MA.GO(evaluationFIS[SU, SLI, SLD]) where SU,SLI,SLD FIS are entries for return rates of both servomotors (MA, MB)
AT_j.MA.DECREASEV
AT_j.MB.GO(evaluationFIS[SU, SLI, SLD]) where SU,SLI,SLD FIS are entries for return rates of both servomotors (MA, MB)
AT_j.MB. DECREASEV
AS. BEGINTASK
AS.RECONOCIMAGE.STATE = (N) where N can have 2 states (YES, NO)
AS.ACTION.TAKEFIGURE
AS.ACTION.AVOID
AS.FIGUREFOUND
AS.ENDMAZE = (M) where M can have 2 states (YES, NO)
AS.ENDTASK

Rules

If AS. BEGINTASK and AN_i.*SU.STATE=OFF* then AN_i.*SU.STATE=ON*
If AS. BEGINTASK and AN_i.*SC.STATE=OFF* then AN_i.*SC.STATE=ON*
If AS. BEGINTASK and AN_i.*SLI.STATE=OFF* then AN_i.*SLI.STATE=ON*
If AS. BEGINTASK and AN_i.*SLD.STATE=OFF* then AN_i.*SLD.STATE=ON*
If AS.ENDTASK and AN_i.*SU.STATE=ON* then AN_i.*SU.STATE=OFF*
If AS.ENDTASK and AN_i.*SC.STATE=ON* then AN_i.*SC.STATE=OFF*
If AS.ENDTASK and AN_i.*SLI.STATE=ON* then AN_i.*SLI.STATE=OFF*
If AS.ENDTASK and AN_i.*SLD.STATE=ON* then AN_i.*SLD.STATE=OFF*
If AS. BEGINTASK and AN_i.*SU.STATE=ON* then AT_j.M.GO(evaluationFIS[SU, SLI, SLD])
If AT_j.M.GO(evaluationFIS[SU, SLI, SLD]) and AN_i.*SU.OBJ*= DETECTS then AN_i.*SC.TIMAGE*
If AN_i.*SC.TIMAGE* then AS.RECONOCIMAGE
If AS.RECONOCIMAGE.STATE = SI then AS.ACTION.TOMAFIG
If AS.RECONOCIMAGE.STATE = NO then AS.ACTION.EVADIR
If AS.ACTION.TAKEFIGURE then AT_j.MA.GO(evaluationFIS[SU, SLI, SLD]) and AT_j.MB.GO(evaluationFIS[SU, SLI, SLD])
If AS.ACTION.AVOID then AT_j.MA.GO(evaluationFIS[SU, SLI, SLD]) and AT_j.MB.GO(evaluationFIS[SU, SLI, SLD])
If AS.ACTION.TAKEFIGURE and AS.ENDMAZE = YES then AS.ENDTASK
If AS.ACTION.TAKEFIGURE and AS.ENDMAZE = NO then AT_j.MA.GO(evaluationFIS[SU, SLI, SLD]) and AT_j.MB.GO(evaluationFIS[SU, SLI, SLD])
If AS.ACTION.AVOID and AS.ENDMAZE = YES then AT_j.MA.GO(evaluationFIS[SU, SLI, SLD]) and AT_j.MB.GO(evaluationFIS[SU, SLI, SLD])

6 Results

We show in Figure 11 the plant and the controller that we are used for achieving the simulation in Simulink of Matlab.

Together with some fuzzy systems which were developed based on work taken as a reference [13][14] and that were the basis for the experimentation with different numbers of rules, parameters of membership functions, and types of membership functions was a good performance of our robot, achieving our goal .

The following fuzzy system has three inputs, two outputs and consists of twenty-seven rules for the inference.

The first input variable of the fuzzy system is the ultrasonic sensor which has three membership functions which are linguistic (close, near, far), as shown in Fig. 13.

The second variable of the fuzzy system is the light sensor that has three membership functions, which are low, middle and high, as shown in Fig. 14.

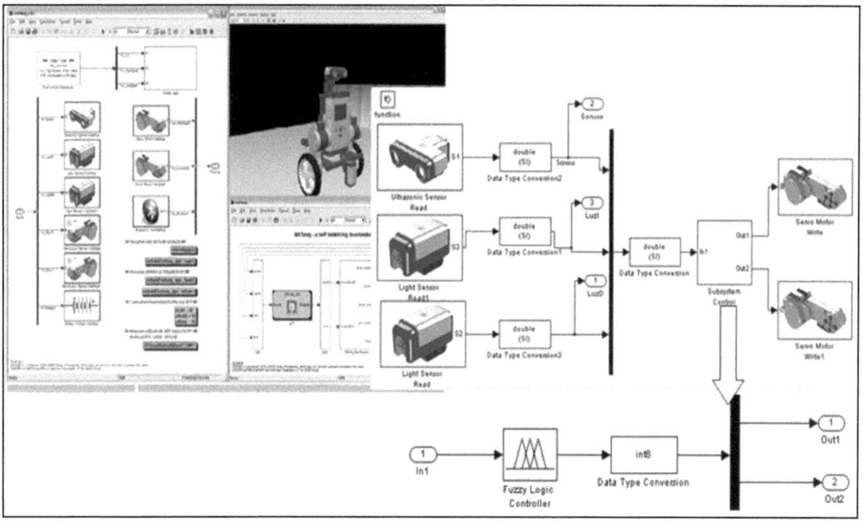

Fig. 11. Plant for simulation.

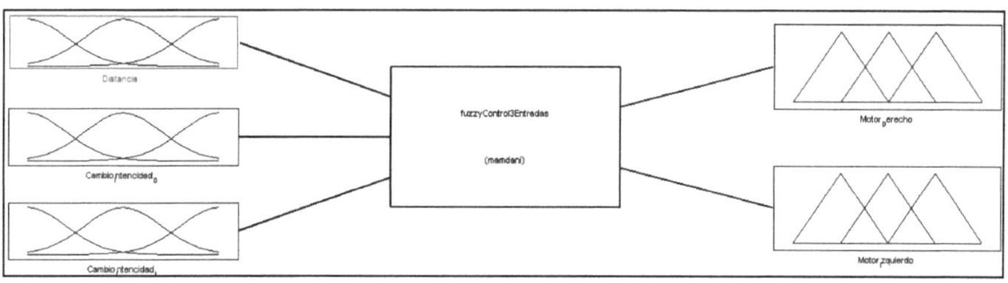

Fig. 12. Shows the structure of the fuzzy system with which it works.

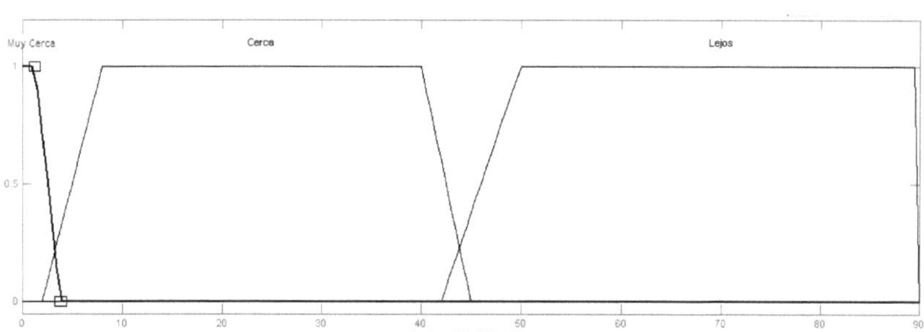

Fig. 13. The Ultrasonic sensor.

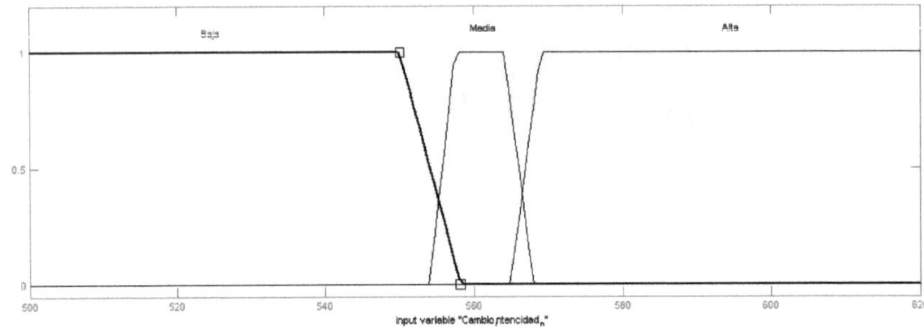

Fig. 14. The left light sensor.

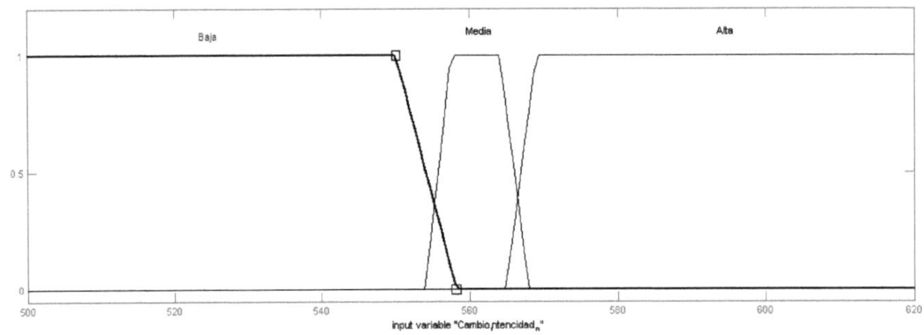

Fig. 15. The right light sensor.

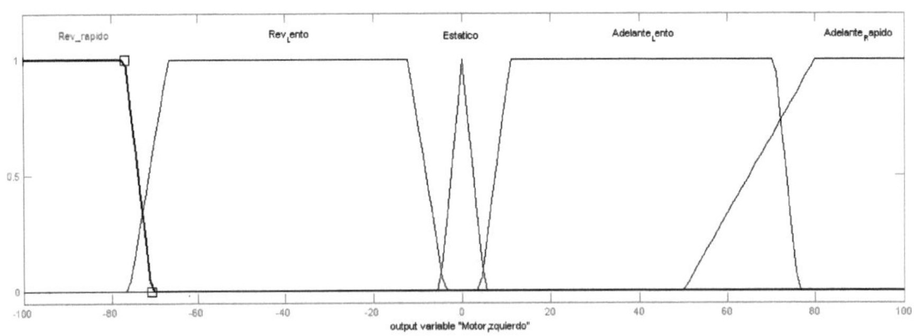

Fig. 16. Motor left.

The third variable of the fuzzy system is the light sensor that has three membership functions, which are low, middle and high, as shown in Fig. 15.

This is the first output variable is the speed of the left engine, has five membership functions as shown in Fig. 16.

This is the second output variable is the speed of the right engine, has five membership functions as shown in Fig. 17.

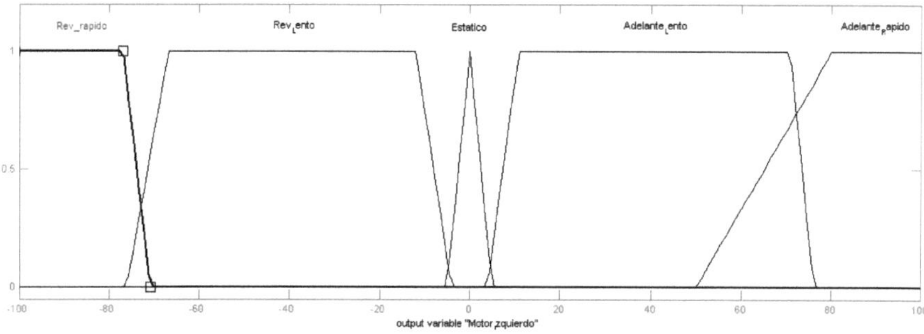

Fig. 17. Motor right.

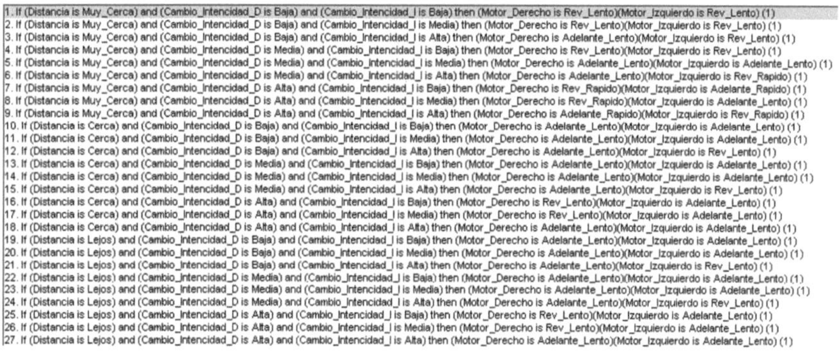

1. If (Distancia is Muy_Cerca) and (Cambio_Intencidad_D is Baja) and (Cambio_Intencidad_I is Baja) then (Motor_Derecho is Rev_Lento)(Motor_Izquierdo is Rev_Lento) (1)
2. If (Distancia is Muy_Cerca) and (Cambio_Intencidad_D is Baja) and (Cambio_Intencidad_I is Media) then (Motor_Derecho is Rev_Lento)(Motor_Izquierdo is Rev_Lento) (1)
3. If (Distancia is Muy_Cerca) and (Cambio_Intencidad_D is Baja) and (Cambio_Intencidad_I is Alta) then (Motor_Derecho is Adelante_Lento)(Motor_Izquierdo is Rev_Lento) (1)
4. If (Distancia is Muy_Cerca) and (Cambio_Intencidad_D is Media) and (Cambio_Intencidad_I is Baja) then (Motor_Derecho is Rev_Lento)(Motor_Izquierdo is Rev_Lento) (1)
5. If (Distancia is Muy_Cerca) and (Cambio_Intencidad_D is Media) and (Cambio_Intencidad_I is Media) then (Motor_Derecho is Adelante_Lento)(Motor_Izquierdo is Adelante_Lento) (1)
6. If (Distancia is Muy_Cerca) and (Cambio_Intencidad_D is Media) and (Cambio_Intencidad_I is Alta) then (Motor_Derecho is Adelante_Lento)(Motor_Izquierdo is Rev_Rapido) (1)
7. If (Distancia is Muy_Cerca) and (Cambio_Intencidad_D is Alta) and (Cambio_Intencidad_I is Baja) then (Motor_Derecho is Rev_Rapido)(Motor_Izquierdo is Adelante_Rapido) (1)
8. If (Distancia is Muy_Cerca) and (Cambio_Intencidad_D is Alta) and (Cambio_Intencidad_I is Media) then (Motor_Derecho is Rev_Rapido)(Motor_Izquierdo is Adelante_Lento) (1)
9. If (Distancia is Muy_Cerca) and (Cambio_Intencidad_D is Alta) and (Cambio_Intencidad_I is Alta) then (Motor_Derecho is Adelante_Rapido)(Motor_Izquierdo is Rev_Rapido) (1)
10. If (Distancia is Cerca) and (Cambio_Intencidad_D is Baja) and (Cambio_Intencidad_I is Baja) then (Motor_Derecho is Adelante_Lento)(Motor_Izquierdo is Adelante_Lento) (1)
11. If (Distancia is Cerca) and (Cambio_Intencidad_D is Baja) and (Cambio_Intencidad_I is Media) then (Motor_Derecho is Adelante_Lento)(Motor_Izquierdo is Adelante_Lento) (1)
12. If (Distancia is Cerca) and (Cambio_Intencidad_D is Baja) and (Cambio_Intencidad_I is Alta) then (Motor_Derecho is Adelante_Lento)(Motor_Izquierdo is Rev_Lento) (1)
13. If (Distancia is Cerca) and (Cambio_Intencidad_D is Media) and (Cambio_Intencidad_I is Baja) then (Motor_Derecho is Adelante_Lento)(Motor_Izquierdo is Adelante_Lento) (1)
14. If (Distancia is Cerca) and (Cambio_Intencidad_D is Media) and (Cambio_Intencidad_I is Media) then (Motor_Derecho is Adelante_Lento)(Motor_Izquierdo is Adelante_Lento) (1)
15. If (Distancia is Cerca) and (Cambio_Intencidad_D is Media) and (Cambio_Intencidad_I is Alta) then (Motor_Derecho is Adelante_Lento)(Motor_Izquierdo is Rev_Lento) (1)
16. If (Distancia is Cerca) and (Cambio_Intencidad_D is Alta) and (Cambio_Intencidad_I is Baja) then (Motor_Derecho is Rev_Lento)(Motor_Izquierdo is Adelante_Lento) (1)
17. If (Distancia is Cerca) and (Cambio_Intencidad_D is Alta) and (Cambio_Intencidad_I is Media) then (Motor_Derecho is Rev_Lento)(Motor_Izquierdo is Adelante_Lento) (1)
18. If (Distancia is Cerca) and (Cambio_Intencidad_D is Alta) and (Cambio_Intencidad_I is Alta) then (Motor_Derecho is Adelante_Lento)(Motor_Izquierdo is Adelante_Lento) (1)
19. If (Distancia is Lejos) and (Cambio_Intencidad_D is Baja) and (Cambio_Intencidad_I is Baja) then (Motor_Derecho is Adelante_Lento)(Motor_Izquierdo is Adelante_Lento) (1)
20. If (Distancia is Lejos) and (Cambio_Intencidad_D is Baja) and (Cambio_Intencidad_I is Media) then (Motor_Derecho is Adelante_Lento)(Motor_Izquierdo is Adelante_Lento) (1)
21. If (Distancia is Lejos) and (Cambio_Intencidad_D is Baja) and (Cambio_Intencidad_I is Alta) then (Motor_Derecho is Adelante_Lento)(Motor_Izquierdo is Rev_Lento) (1)
22. If (Distancia is Lejos) and (Cambio_Intencidad_D is Media) and (Cambio_Intencidad_I is Baja) then (Motor_Derecho is Adelante_Lento)(Motor_Izquierdo is Adelante_Lento) (1)
23. If (Distancia is Lejos) and (Cambio_Intencidad_D is Media) and (Cambio_Intencidad_I is Media) then (Motor_Derecho is Adelante_Lento)(Motor_Izquierdo is Adelante_Lento) (1)
24. If (Distancia is Lejos) and (Cambio_Intencidad_D is Media) and (Cambio_Intencidad_I is Alta) then (Motor_Derecho is Adelante_Lento)(Motor_Izquierdo is Rev_Lento) (1)
25. If (Distancia is Lejos) and (Cambio_Intencidad_D is Alta) and (Cambio_Intencidad_I is Baja) then (Motor_Derecho is Rev_Lento)(Motor_Izquierdo is Adelante_Lento) (1)
26. If (Distancia is Lejos) and (Cambio_Intencidad_D is Alta) and (Cambio_Intencidad_I is Media) then (Motor_Derecho is Rev_Lento)(Motor_Izquierdo is Adelante_Lento) (1)
27. If (Distancia is Lejos) and (Cambio_Intencidad_D is Alta) and (Cambio_Intencidad_I is Alta) then (Motor_Derecho is Adelante_Lento)(Motor_Izquierdo is Adelante_Lento) (1)

Fig. 18. Rules of the fuzzy system (FIS).

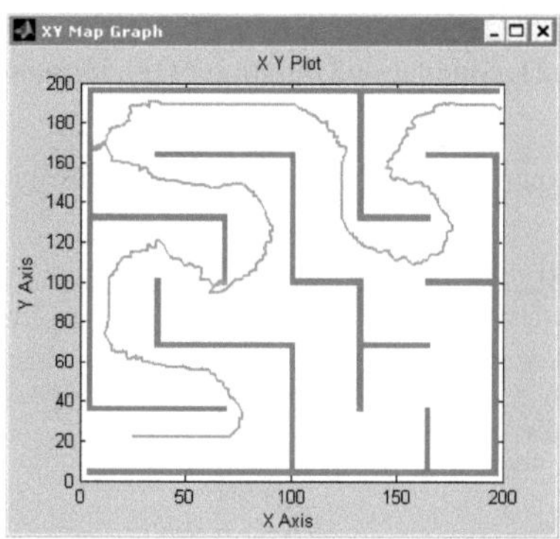

Fig. 19. Simulation1 duration of 01:03 minutes.

In Fig. 18 we show the rules that are used in our system, which are 10 fuzzy rules.

An example of the results simulating this fuzzy system is shown in the figures 19, 20 and 21, in which we see that as our robot moves forward in the path of the maze to find out achieved or failing not find it.

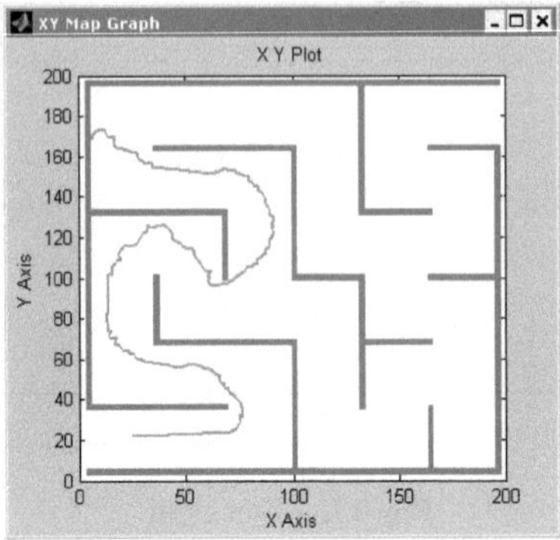

Fig. 20. Simulation2 duration of 01:30 minutes.

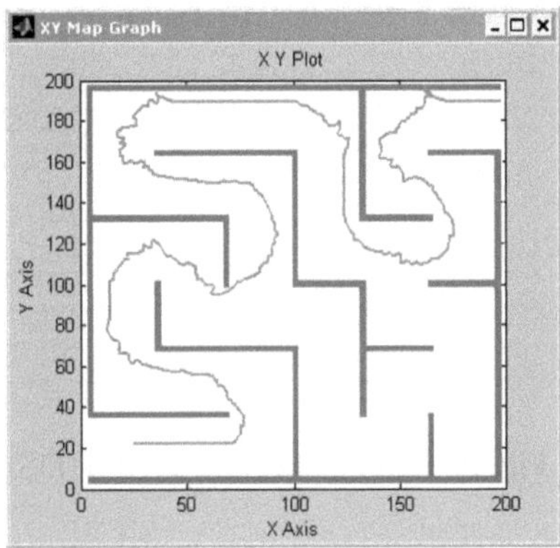

Fig. 21. Simulation3 duration of 00:58 minutes.

Table 1. Experiments shows our fuzzy systems with 27 rules.

Number of experiment	Duration/minutes	Exit the maze
1	01:03	Yes
2	01:30	No
3	00:58	Yes
4	00:50	No
5	00:53	Yes
6	01:17	No
7	00:57	No
8	00:58	Yes
9	01:01	Yes
10	00:60	Yes
11	01:35	No
12	01:46	No

Note: It is highlighted in yellow the best time to exit the maze

Table 2. A comparison between simulations of the work of reference[13][14].

Number of experiment	Duration/minutes	Exit the maze
1	00:25	No
2	00:25	No
3	00:25	No
4	00:25	No
5	00:25	No
6	00:25	No
7	00:25	No
8	00:25	No
9	01:00	Yes
10	00:59	Yes
11	00:39	No
12	01:00	Yes

Note: It is highlighted in blue the best time to exit the maze

Above are two tables in which present the results obtained in experiments, where it can be noticed that until now we have a better time to exit the maze, in both tables highlight the best time of departure of the experiments; experiments have been made in regard to equality in computer equipment and conditions maze. Table 1 presents the results of twelve experiments on the duration that each one had escaped, and whether or not in the same way in Table 2 presents the results of the work taken as a reference[13][14].

In figure 22 we show the results presented above for a better understanding, which can be seen in the upper part the time it takes for each experiment out of the

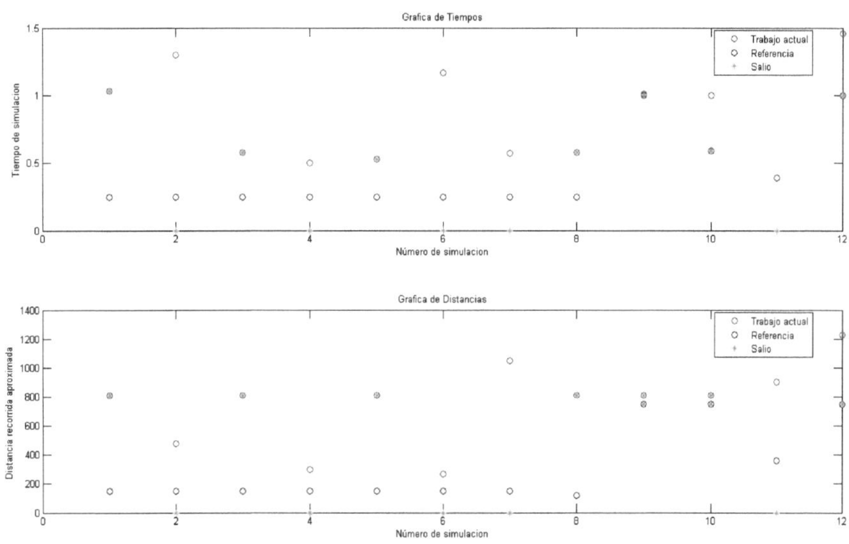

Fig. 22. Graphic comparison between the work of reference [13][14] and the current job.

maze at the bottom and the approximate distance traveled to each robot in the experiment, the red dots are the results of our work, the blue dots are the results of the reference points and finally green means yes or no out of the labyrinth in this experiment.

Thcrc arc occasions whcn our results are very similar and therefore only seems to be the green dot that tells us whether or not we go out of the labyrinth that is the case of experiment number nine in the graph of time in which both works coincided in time output and both were successful in finding the exit of the maze.

Now we show other experiments with a fuzzy system with different number of fuzzy rules, membership functions and parameters of membership functions. It is noteworthy that although this is not a fuzzy system which is optimized for this any optimization method can be used.

The following fuzzy system in Fig.23 has three inputs, two outputs and consists of ten rules to make the inference.

The first input variable of the fuzzy system is the ultrasonic sensor which has three membership functions which are linguistic (close, near, far), as shown in Fig. 24.

The second variable of the fuzzy system is the light sensor that has two functions that are membership free and wall, as shown in Fig. 25.

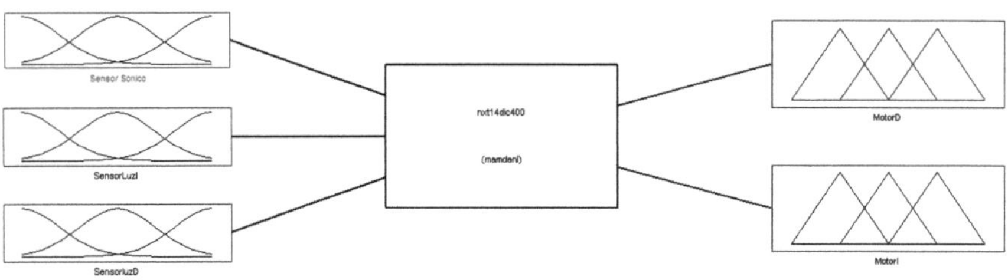

Fig. 23. Shows the fuzzy system that works for the simulations.

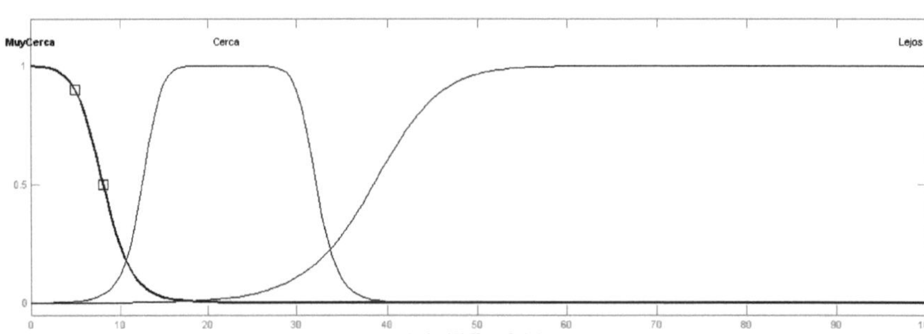

Fig. 24. The ultrasonic sensor.

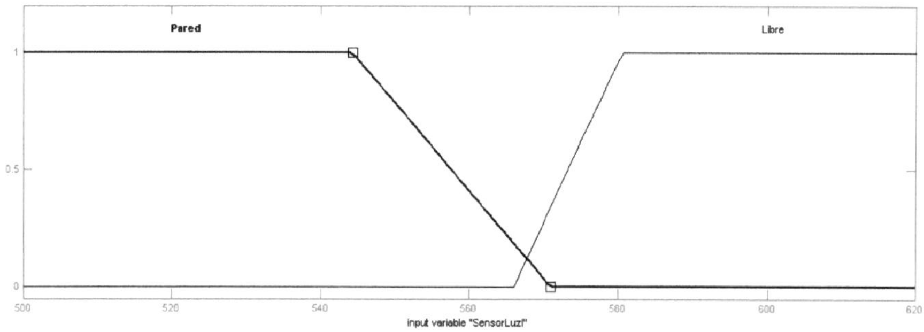

Fig. 25. The left light sensor.

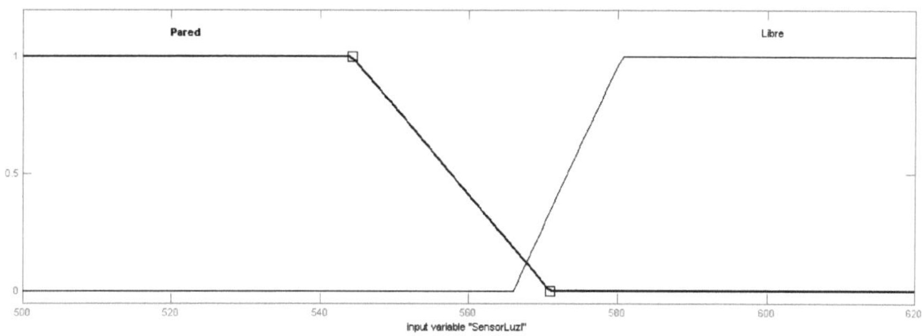

Fig. 26. The right light sensor.

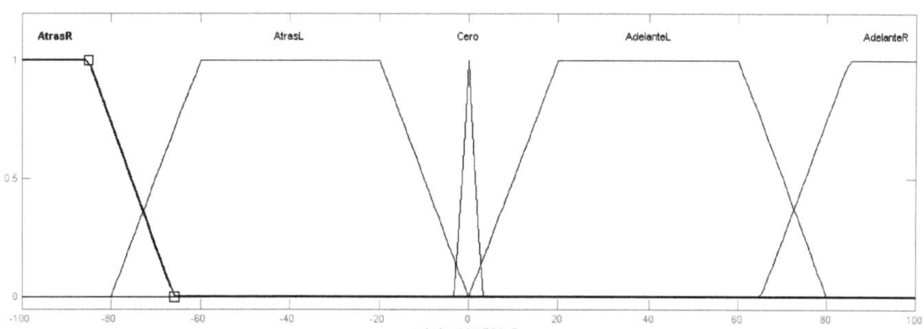

Fig. 27. Motor left.

The third variable of the fuzzy system is the light sensor that has two functions that are membership free and wall, as shown in Fig. 26.

This is the first output variable is the speed of the left engine, has five membership functions as shown in Fig. 27.

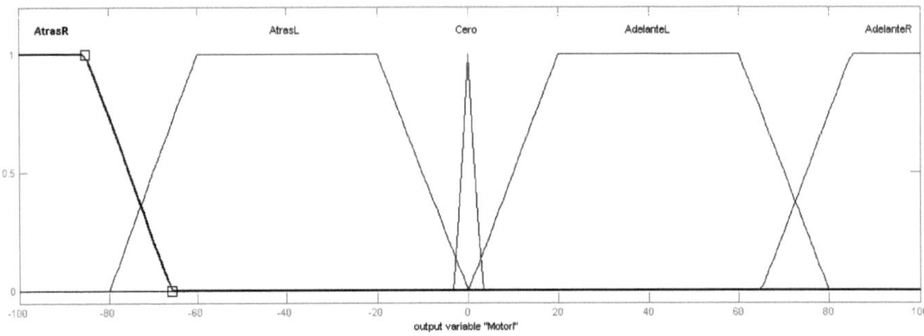

Fig. 28. Motor right.

1. If (SensorSonico is Lejos) and (SensorLuzI is Pared) and (SensorluzD is Pared) then (MotorD is AdelanteR)(MotorI is AdelanteR) (1)
2. If (SensorSonico is Cerca) and (SensorLuzI is Pared) and (SensorluzD is Pared) then (MotorD is AdelanteL)(MotorI is AdelanteL) (1)
3. If (SensorSonico is MuyCerca) and (SensorLuzI is Pared) and (SensorluzD is Pared) then (MotorD is AdelanteL)(MotorI is AdelanteL) (1)
4. If (SensorSonico is Lejos) and (SensorLuzI is Libre) and (SensorluzD is Pared) then (MotorD is AdelanteR)(MotorI is AtrasL) (1)
5. If (SensorSonico is Cerca) and (SensorLuzI is Pared) and (SensorluzD is Libre) then (MotorD is AtrasL)(MotorI is AdelanteR) (1)
6. If (SensorSonico is MuyCerca) and (SensorLuzI is Pared) and (SensorluzD is Libre) then (MotorD is AtrasL)(MotorI is AdelanteR) (1)
7. If (SensorSonico is Lejos) and (SensorLuzI is Pared) and (SensorluzD is Libre) then (MotorD is AtrasL)(MotorI is AdelanteR) (1)
8. If (SensorSonico is Cerca) and (SensorLuzI is Libre) and (SensorluzD is Libre) then (MotorD is AdelanteR)(MotorI is Cero) (1)
9. If (SensorSonico is Lejos) and (SensorluzD is Pared) then (MotorD is AdelanteL)(MotorI is AdelanteL) (1)
10. If (SensorSonico is MuyCerca) and (SensorLuzI is Libre) and (SensorluzD is Pared) then (MotorD is AdelanteR)(MotorI is AtrasL) (1)

Fig. 29. Rules of the fuzzy system (FIS).

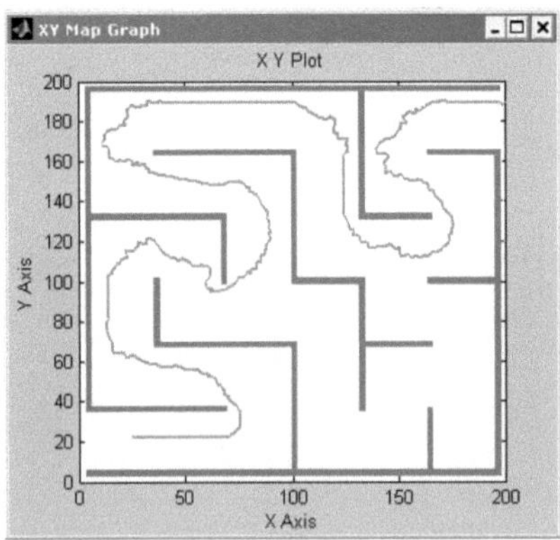

Fig. 30. Simulation4 duration of 00:49 minutes.

This is the second output variable is the speed of the right engine, has five membership functions as shown in Fig. 28.

In Fig. 29 we show the rules that are used in the fuzzy system, which are 10 fuzzy rules.

The results simulating this fuzzy system are shown in the figures 30, 31 and 32, in which we see as our robot moves forward in the path of the maze to find out achieved or failing not find it.

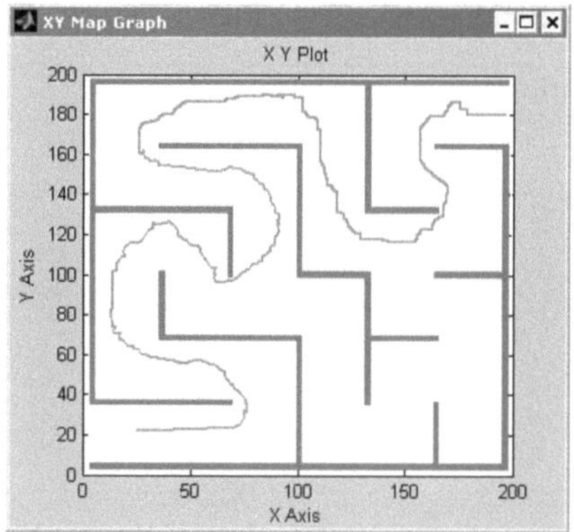

Fig. 31. Simulation5 duration of 00:45 minutes.

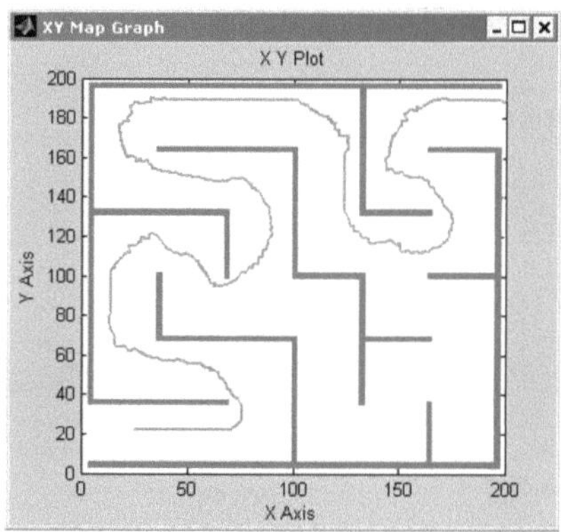

Fig. 32. Simulation6 duration of 00:47 minutes.

Below are two tables corresponding to the experiments with the fuzzy systems described above, we present the results obtained in experiments, where it can be noticed that until now we have a better time to exit the maze with respect to reference work [13][14], in both tables highlight the best time of departure of the experiments, the experiments have sought to make equal in terms of computer equipment and conditions maze. In Table 3 presents the results of twelve experiments on the duration that each one had escaped, and whether or not in the same way in Table 4 presents the results of the work taken as a reference.

In figure 33 we show the results presented above for a better understanding where you can appreciate the time it takes for each experiment out of the labyrinth, the red dots are the results of our work, the blue dots are the results of reference and finally the green means yes or no out of the labyrinth in this experiment.

Table 3. Experiments shows our fuzzy systems with 10 rules.

Number of Simulation	Duration/minutes	Exit the maze
1	00:56	Yes
2	00:51	Yes
3	00:55	Yes
4	00:49	Yes
5	00:45	Yes
6	00:47	Yes
7	00:47	Yes
8	00:50	Yes
9	00:51	Yes

Note: It is highlighted in yellow the best time to exit the maze.

Table 4. A comparison between simulations of reference work[13][14].

Number of Simulation	Duration/minutes	Exit the maze
1	00:55	Yes
2	00:51	Yes
3	00:55	Yes
4	00:54	Yes
5	00:50	Yes
6	00:46	Yes
7	00:50	Yes
8	00:52	Yes
9	00:55	Yes

Note: It is highlighted in blue the best time to exit the maze

There are occasions when our results are very similar and therefore only seems to be the green dot that tells us whether or not we go out of the labyrinth that is the case of experiment number nine in the graph of time in which both works coincided in time output and both were successful in finding the exit of the maze. Now in terms of performance and implementation of the System of Multi-agent simulations are presented below as an example of the results of the SMA. Where we began working with the best system of 10 fuzzy rules since in all simulations find the exit of the labyrinth for which we believe our best option despite being aware that the movements were not performed as best as they are very prone to movement.

Table 5 presents the results obtained in the experiments, where it can be noticed that we have obtained a better time to leave the maze in the implementation of MAS.

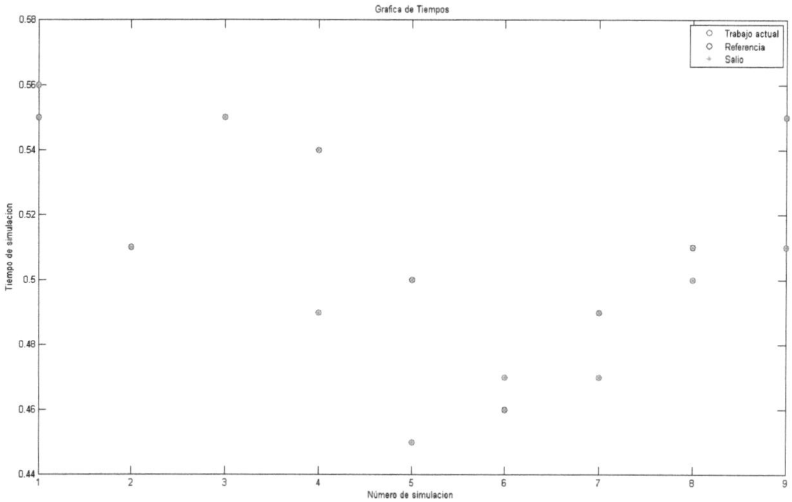

Fig. 33. Graphic comparison between the work of reference and the current job.

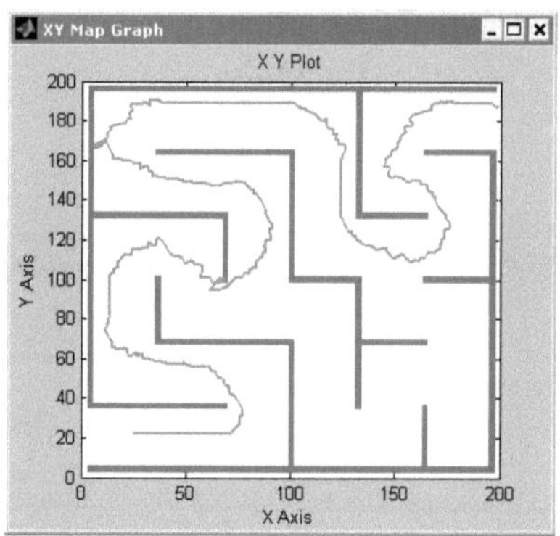

Fig. 34. Simulation4 duration of 00:50 minutes.

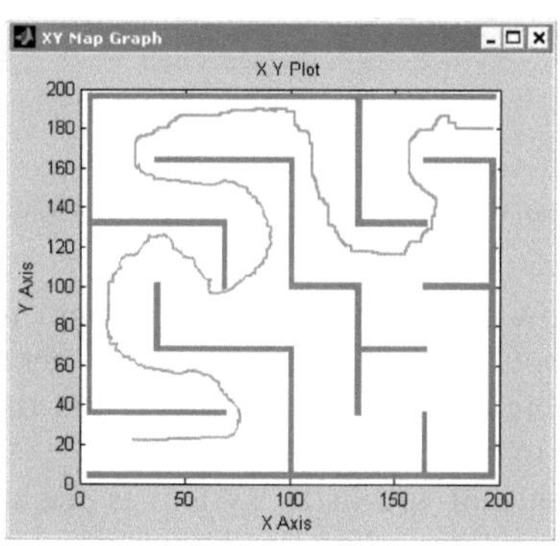

Fig. 35. Simulation5 duration of 00:48 minutes.

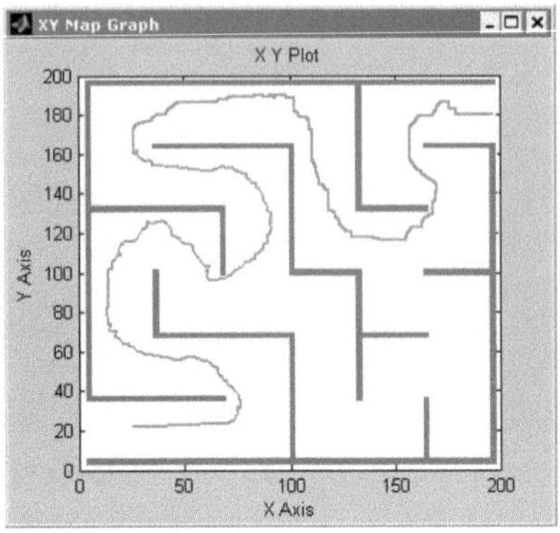

Fig. 36. Simulation6 duration of 00:50 minutes.

Table 5. A comparison between different simulations of work.

Number of Simulation	Duration/minutes	Exit the maze
1	00:50	Yes
2	00:50	Yes
3	00:51	Yes
4	00:50	Yes
5	00:48	Yes
6	00:50	Yes
7	00:53	Yes
8	00:51	Yes
9	00:52	Yes

Note: Is highlighted in yellow the best time to exit the maze.

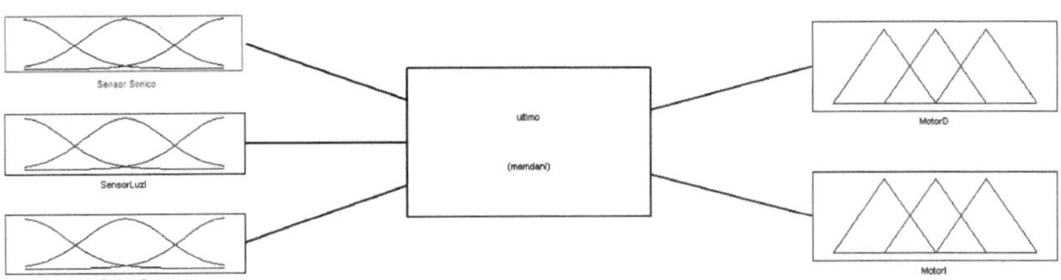

Fig. 37. Shows the fuzzy system that works for the simulations.

Now we consider how to modify the fuzzy system to produce more gentle movements of the robot, using another type of membership functions and parameters. The following fuzzy system in Fig.37 has three inputs, two outputs and consists of ten rules to make the inference.

The first input variable of the fuzzy system is the ultrasonic sensor which has three membership functions which are linguistic (close, near, far), as shown in Fig. 38.

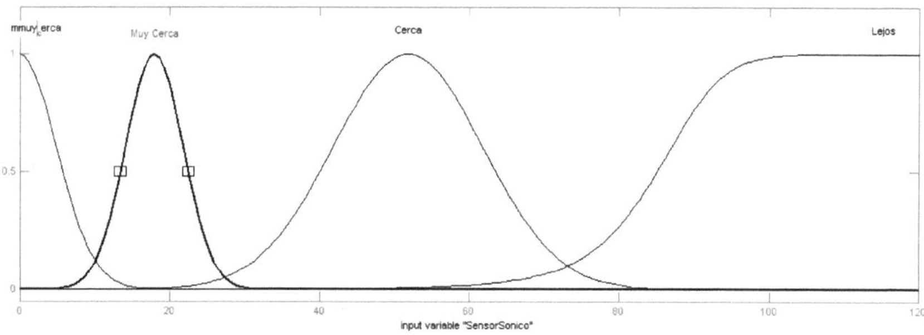

Fig. 38. The ultrasonic sensor.

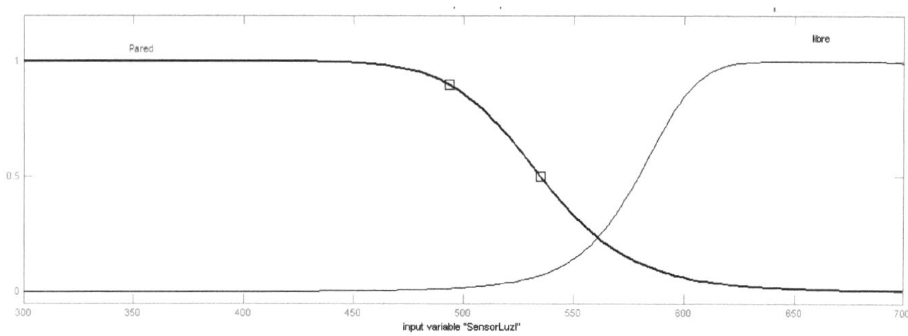

Fig. 39. The left light sensor.

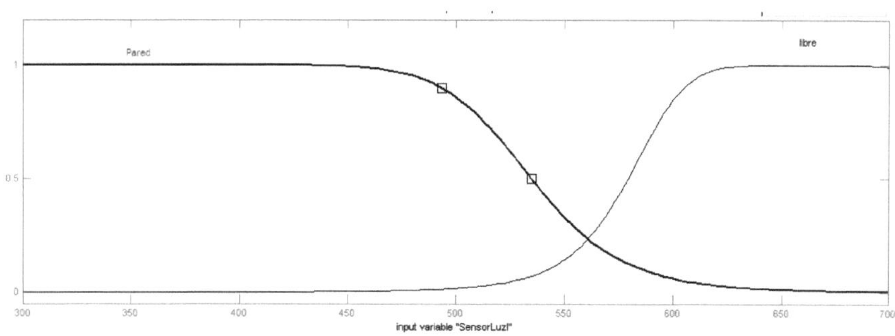

Fig. 40. The right light sensor.

The second variable of the fuzzy system is the light sensor that has two functions that are membership free and wall, as shown in Fig. 39.

The third variable of the fuzzy system is the light sensor that has two functions that are membership free and wall, as shown in Fig. 40.

This is the first output variable is the speed of the left engine, has five membership functions as shown in Fig. 41.

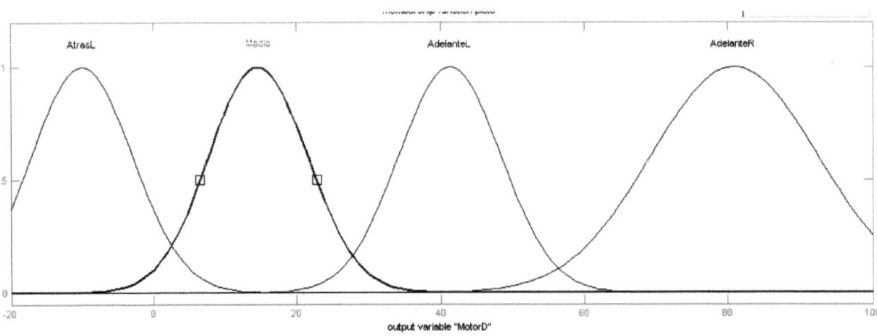

Fig. 41. Left motor.

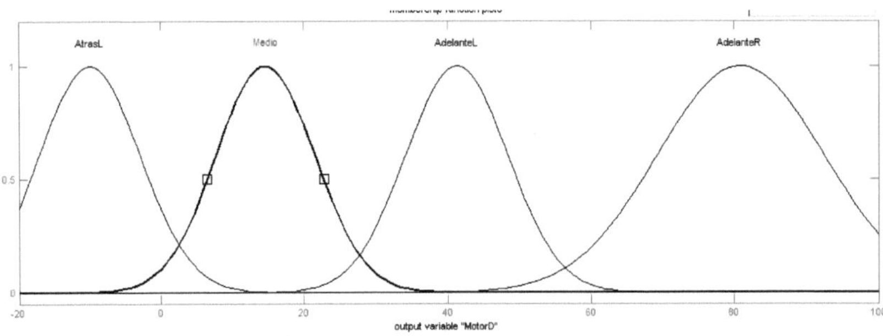

Fig. 42. Right motor.

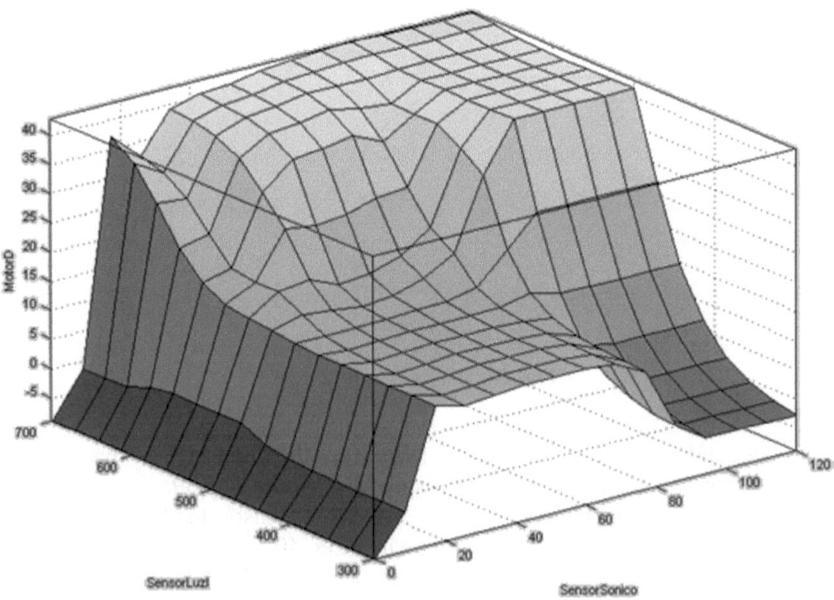

Fig. 43. Surface of fuzzy system.

This is the second output variable is the speed of the right engine, has five membership functions as shown in Fig. 42.

In Fig. 43 we see the surface of the behavior of our fuzzy systems where it is seen that although not all have a smooth behavior.

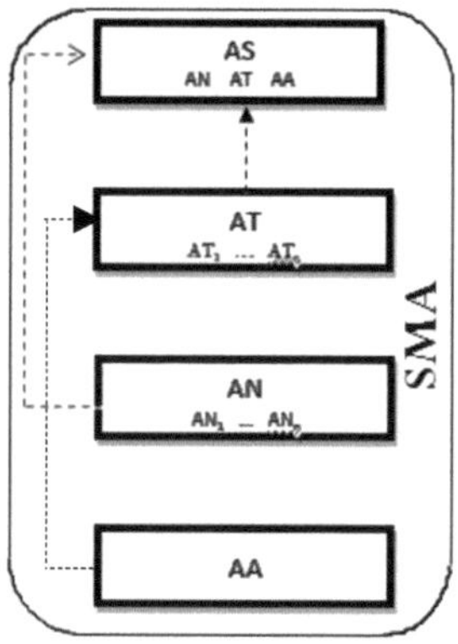

Fig. 44. New scheme of Multi-Agent System.

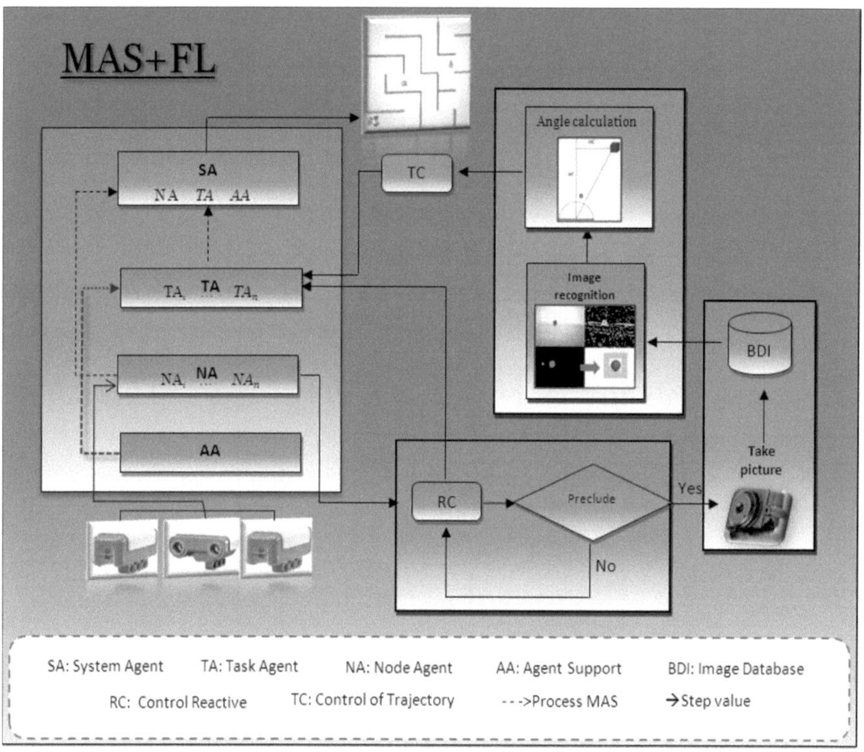

Fig. 45. New architecture of the complete system.

Having modified and improved the fuzzy systems, we proceeded with a slight modification to our Multi-Agent System, which consists in adding a new agent which has the task of giving a second option in case our robot gets stuck somewhere in the world, for example corners in which our sensors cannot do some action, then come into action. The new agent called Support Agent enters into action as I mentioned in the case that the sensor values do not become active some rule of fuzzy systems that we throw the speed of the drive to get out of where we are stuck with only works with the task agent (TA) for which our multi-agent system would be as shown in Fig. 44.

Integrating it to our general scheme we would be as shown in Fig 45.

Now with these changes we had better control of movements of the robot in the search for the exit of the maze, which can be seen in the graphs of the following experiments.

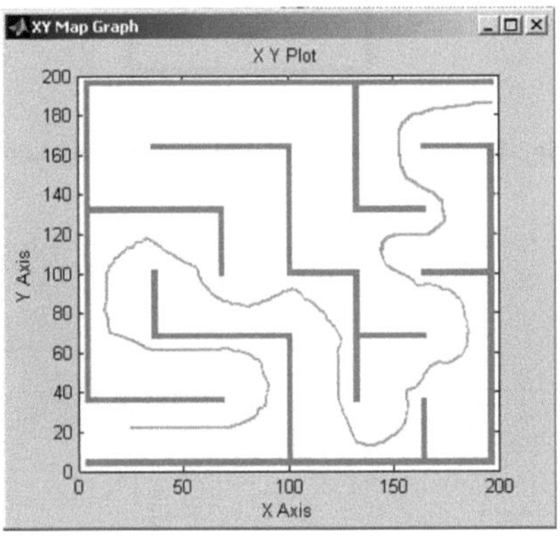

Fig. 46. Simulation2 duration of 00:55 minutes.

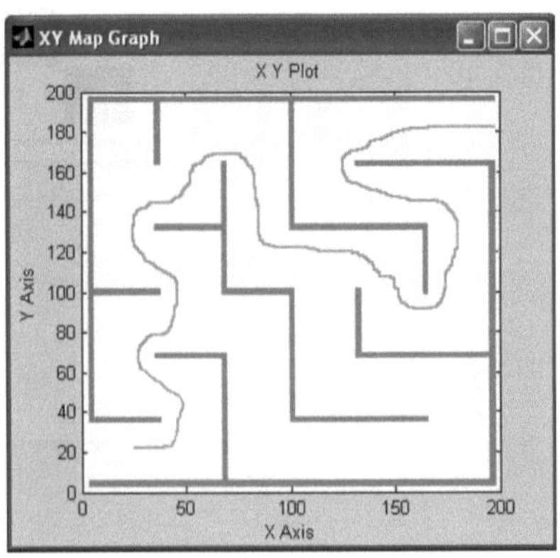

Fig. 47. Simulation5 duration of 00:47 minutes.

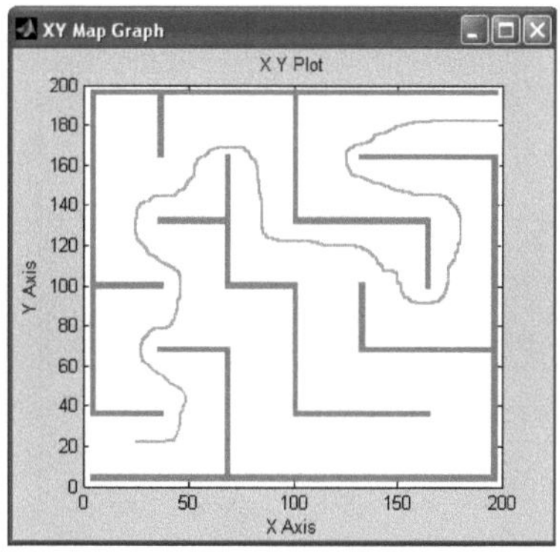

Fig. 48. Simulation7 duration of 00:47 minutes.

Table 6. A comparison between the simulations of our work.

Number of Simulation	Duration/minutes	Exit the maze
1	00:50	Yes
2	00:55	Yes
3	00:51	Yes
4	00:53	Yes
5	00:47	Yes
6	00:50	Yes
7	00:49	Yes
8	00:53	Yes
9	00:47	Yes

Note: Is highlighted in yellow the best time to exit the maze.

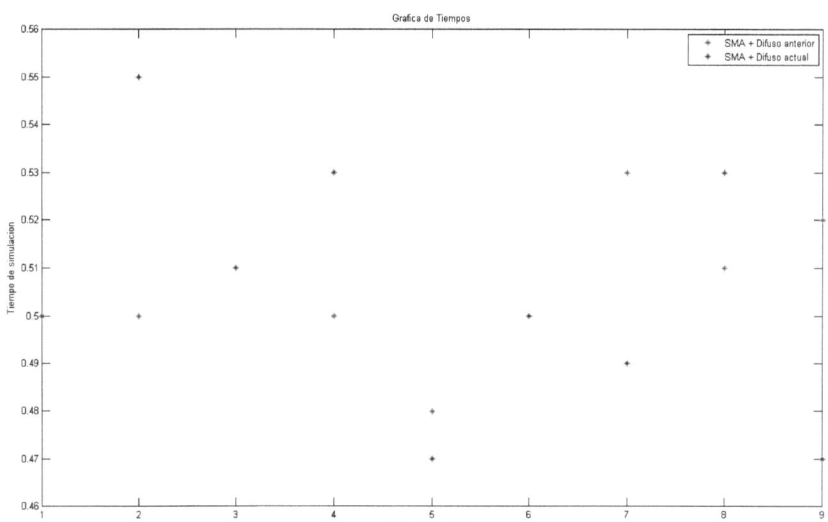

Fig. 49. Graphic comparison between the original and current work.

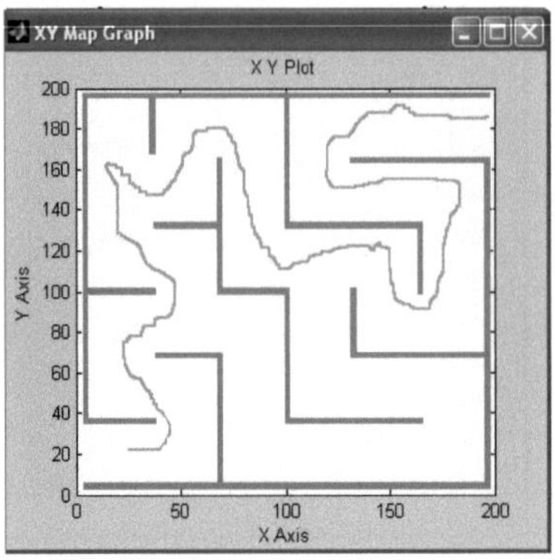

Fig. 50. Simulation1 with maze backwards duration of 00:50 minutes.

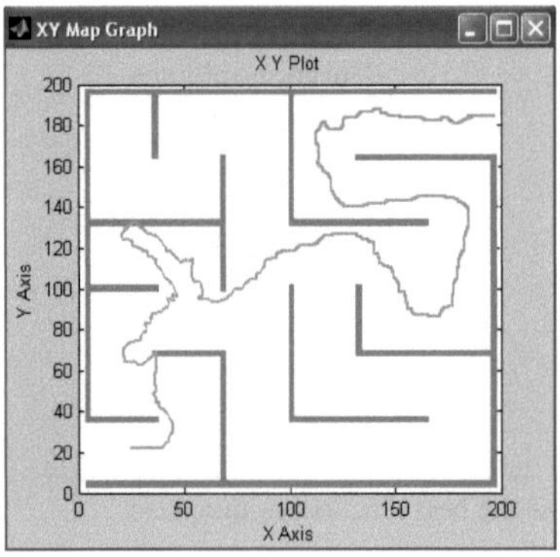

Fig. 51. Simulation4 with more obstacles duration of 00:56 minutes.

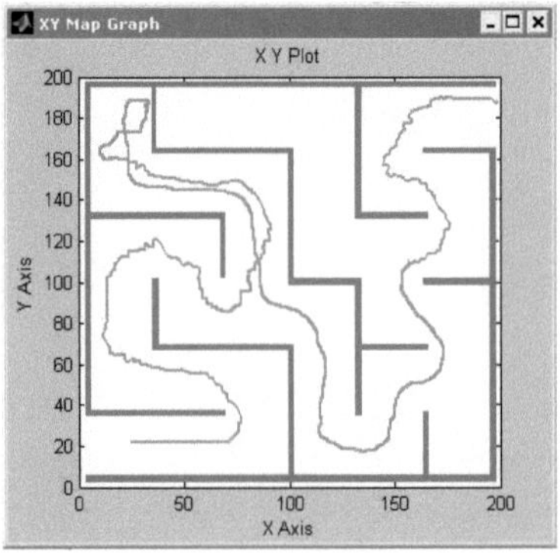

Fig. 52. Simulation7 only one way duration of 01:05 minutes.

Table 7. Simulation results with the modified maze.

Number of Simulation	Duration/minutes	Exit the maze
1	00:50	Yes
2	**00:49**	**Yes**
3	00:51	Yes
4	00:56	Yes
5	00:56	Yes
6	**00:55**	**Yes**
7	*01:05*	*Yes*
8	*01:00*	*Yes*
9	***00:59***	***Yes***

Note: The color difference in the three modified maze and in bold the best times of any changes.

Now we present the table corresponding to the time of the simulation experiments with the System and Multi-Agent System fuzzy above.

In table 7 we see the results of the mazes with which to verify the experience modification that was applied.

In Table 7 above we can see the times of the simulation experiments in which we can perceive three colors which we refer to the three types of changes that were made to the question of mazes to test our system, the modifications we made were:

1. Flip the maze completely to place it opposite the entrance and exit of the maze.
2. Add obstacles to the maze at different locations.
3. Allow only one way out so forcing the robot to search.

7 Conclusions

The development of intelligent applications is varied, in this particular case, the implementation of multi-agent systems is complex because it consists of several agents, whit the incorporation of other techniques, making it even more complex.

One of the main objectives of this research initially was to open a new area of intelligent agents using fuzzy logic, which hitherto has not been given the approach that is intended here, which gives a degree of utilization of this paradigm of intelligent agents, but more research is needed in order to defend this idea as innovative.

Considering the non-optimized fuzzy systems, we have shown better results than were obtained in the work we take as a reference [14] for what we have achieved a good percentage of the overall project objectives and to improve not only the times but we have improved how it moves the robot by making our world more soft and therefore earns a bit of time helping to improve.

And we gave another level to our Multi-Agent System as it not only succeeded in getting out of the labyrinth with which we have been working but we show results where the maze was changed in different ways and still finds the exit and not in a very long time. Indeed the average that has been obtained in the work of reference and is even a bit less here, so we are happy with the results we have achieved over not satisfied because we still have work to resolve because we have not yet optimized the system, with the hope that optimization can further improve the results obtained previously and perhaps even lower computation times are achieved.

References

[1] Alanis, A.: Contribution to the design of Fault Tolerant Distributed Systems for industrial control: proposal for a new paradigm based on Intelligent Agents, PhD, Universidad Politecnica de Valencia (November 2007)

[2] Alanis, A., López, M., Muñoz, R., Pulido, M., Martínez, P., Beltrán, M.: Multi-agent System for a Lego NXT robot to access the collaboration object 1.8 °. In: National Congress of Electrical and Electronics Engineering of the Mayab conieem 2008 Instituto Tecnológico de Mérida, of Merida, Yucatan, Mexico, April 21-25 (2008)

[3] Alanis, A., López, M., Muñoz, R., Pulido, M., Martínez, P., Beltrán, M.: Multi-agent system for a Lego NXT robot to access the collaboration of objects 1, XXIX International Congress of Engineering in electronics, electro 2008, Instituto Tecnológico de Chihuahua, division of graduate studies and research, October 29-31, Chihuahua, Chih. Mexico (2008)

[4] Alanis, A., López, M., Parra, B., Serrano, M., Ayala, E., Solano, C.: Multi-agent system for search and object recognition using vision, in a Lego NXT robot 1.8 °. National Congress of Electrical and Electronics Engineering of the Mayab conieem 2008 Instituto Tecnológico de Mérida, of Merida, Yucatan, Mexico, April 21-25 (2008)

[5] Alanis, A., López, M., Parra, B., Serrano, M., Ayala, E., Solano, C.: Multi-agent system for search and object recognition using vision. In: A Lego NXT robot 1, XXIX International Congress of Engineering in electronics, electro 2008 Instituto Tecnologico de Chihuahua, division of graduate studies and research, Chihuahua, Chih. Mexico, October 29–31 (2008)

[6] Aspiazu shock G.: Consejo Superior de Investigaciones Científicas, Madrid. Neural networks and genetic algorithms (2000)

[7] Azam, F.: Biologically Inspired Modular Neural Networks. Electrical and Computer Engineering, Blacksburg, Virginia (2000)

[8] Galván, R., Monsivais, A.: Resolution of mazes with Lego Mindstorms NXT with LabVIEW 7.1, thesis work ITT

[9] Jang, J.-S.R., Sun, C.-T., Mizutani, E.: Neuro-Fuzzy and Soft Computing for computational approach to learning and machine intelligence. Prentice Hall, New Jersey (1997)

[10] Jang, S.J., Mizutani, J.: Neuro-Fuzzy and soft computing: a computational approach to learning and machine intelligence. Prentice-Hall, Englewood Cliffs (1997)

[11] López, A.: Program Development in Matlab, Matlab... no. Bulletin of the Cuban Society of Mathematics and Computer Science 2(2), 105–112 (2004)

[12] Maček, K., Petrović, I., Siegwart, R.: A control method for stable and smooth path following of mobile robots. In: Proceedings of the 2nd European Conference on Mobile Robots - ECMR 2005, Ancona, Italy, September 7-10, pp. 128–133 (2005)

[13] Melendez, A.: Control and monitoring reagent for a mobile robot using fuzzy logic, thesis work, ITT (August 2008)

[14] Melendez, A., Castillo, O., Soria, J.: Reactive Control of a Mobile Robot in a Distributed Environment Using Fuzzy Logic. In: NAFIS (2008)

[15] Morgan, D.P., Scofield, C.L.: Neural Networks and Speech Processing. Kluwer Academic Publishers, Dordrecht (1991)

[16] Norving, P., Russell, S.: Artificial intelligence a modern approach. Prentice-Hall, Australia (1996)

[17] Or Etzioni, Lesh N., Segal, R.: Bulding for Softbots UNIX (preliminary report). Tech. Report 93-09-01. Univ. of Washington, Seattle (1993)

[18] Cohen, P.R., et al.: An Open Agent Architecture, working Notes of the AAAI Spring symp.: Software Agent, pp. 1–8. AAAI Press, Cambridge (1994)

[19] Russell, S., Norvig, P.: Intelligent Agent. In: Artificial Intelligence to Modern Aproach. Pretence artificial Hall series in intelligence, pp. 31–52 (1994)

[20] Sánchez, P.: Voice and image recognition by neural networks to guide a robot, thesis work ITT (October 2008)

[21] Sánchez, P., Melin, P., López, M.A.: Hybrid Neural-Based Guiding System for Mobile Robots. In: NAFIS 2008 (2008)

[22] Suárez, J.: Location and tracking of trajectories with robots Wayfarer in natural surroundings, thesis work Universidad Complutense de Madrid

[23] University of Vigo, Department of Computer Science, Problems of search and optimization (October 2004)

[24] Zadeh, L.A.: Outline of a new approach to the analysis of complex systems and decision processes. IEEE Transactions on Systems, Man, and Cybernetics 31(6), 891–901 (2001)

[25] Zadeh, L.A.: The concept of a linguistic variable and its application to approximate reasoning, Parts 1, 2, and 3. Information Sciences (1975)

[26] Intelligent Autonomous Robots are the new generation (September 03, 2008), http://www.tendencias21.net/Los-Robots-Inteligentes-Autonomos-son-la-nueva-generacion_a744.html

[27] Introduction to intelligent control (September 03, 2008), http://isa.umh.es/isa/es/asignaturas/cas/TRANSPCI.PDF

[28] What is Simulink? (September 04, 2008), http://voltio.ujaen.es/jaguilar/matlab/Manual20Matlab_Simulink/manual%%20simulink/SIM_01%20-%20Que_%20es.htm

[29] Simulation of a fuzzy system for controlling engine speed of a CD (September 08, 2008), http://www.uaem.mx/cicos/memorias/3ercic2004/Articulos/articulo3.pdf

[30] fuzzy systems (September 08, 2008), http://ants.dif.um.es/staff/juanbot/ml/files/20022003/fuzzy.pdf

[31] Control of local mobile robots based on statistical methods and genetic algorithms (October 01, 2008), http://www.dccia.ua.es/ria/artics/caepia97.pdf

[32] The NXT Bluetooth C + + library (October 06, 2008), http://www.norgesgade14.dk/bluetoothlibrary.php

[33] Embedded Coder Robot NXT (October 06, 2008), http://www.mathworks.com/matlabcentral/fileexchange/loadFile.do?objectId=13399

[34] Tools for programming free agents (October 13, 2008), http://torio.unileon.es/~vmo/pubs/waf00.pdf
[35] Analysis and design of reinforced learning agents (October 13, 2008), http://www.itcm.edu.mx/mccc06/publicaciones/Analisisydise%F1odeagentes.pdf
[36] An architecture for mobile robots small (October 13, 2008), http://igarrido.vtrbandaancha.net/papers/gbot_arquitectura.pdf
[37] Class diagrams, UML artifacts (October 22, 2008), http://www.vico.org/aRecursosPrivats/UML_TRAD/talleres/mapas/UMLTRAD_101A/LinkedDocuments/UML_diagClases.pdf
[38] Maldonado, O.: Republic of Knowledge, Computer Vision, http://www.depi.itch.edu.mx/apacheco/expo/html/ai11/vision.html#page3
[39] Salinas, R.: Neural Network Architecture Parametric Face Recognition, University of Santiago de Chile, http://cabierta.uchile.cl/revista/17/articulos/pdf/paper4.pdf
[40] http://www.fipa.org/specifications/lifecycle.html
[41] http://www.fipa.org/specifications/identifiers.html

Particle Swarm Optimization Applied to the Design of Type-1 and Type-2 Fuzzy Controllers for an Autonomous Mobile Robot

Ricardo Martínez-Marroquín, Oscar Castillo, and José Soria

Tijuana Institute of Technology, Tijuana México
e-mail: ocastillo@hafsamx.org

Abstract. In this paper we describe the application of a Particle Swarm Optimization (PSO) algorithm as a method of optimization for membership functions' parameters of a fuzzy logic controller (FLC) in order to find the optimal intelligent controller for an Autonomous Wheeled Mobile Robot. Simulations results show that PSO is able to optimize the tipe-1 and type-2 FLCs for this application.

1 Introduction

Nowadays, fuzzy logic is one of the most used methods of computational intelligence and with the best future; this is possible thanks to the efficiency and simplicity of Fuzzy Systems since they use linguistic terms similar to those that human beings use.

The complexity in developing fuzzy systems can be found at the time of deciding which are the best parameters of the membership functions, the number of rules or even the best granularity that could give us the best solution for the problem that we want to solve.

A solution for the above mentioned problem is the application of bio-inspired algorithms for the optimization of fuzzy systems. Optimization algorithms can be a useful tool due to their capabilities of solving nonlinear problems, well-constrained or even NP-hard problems. Among the most used optimization methods we can find: Genetic Algorithms (GA), Ant Colony Optimization (ACO), Particle Swarm Optimization (PSO), etc.

This paper describes the application of PSO as an optimization method of the parameters of the membership functions of the FLC in order to find the best possible intelligent controller for an Autonomous Wheeled Mobile Robot.

This paper is organized as follows: Section 2 shows the basic concepts of PSO and a description of the *gbest* algorithm, which is the technique that was applied for optimization. Section 3 presents the problem statement and the dynamic and kinematic model of the unicycle mobile robot. Section 4 shows the fuzzy logic controller proposed and in Section 5 the development of the optimization method is described. In Section 6 the simulation results are shown. Finally, Section 7 shows the Conclusions.

P. Melin et al. (Eds.): Bio-inspired Hybrid Intelligent Systems, SCI 256, pp. 247–262.
springerlink.com

2 *gbest* PSO Algorithm

PSO is a stochastic optimization technique based on population inspired in the social behavior of big masses of birds, fish or bees.

In PSO, the potential solutions (called particles), fly trough the space problem, where the less optimum particles fly to optimum particles, doing this iteratively until all the particles converge at the same point (solution). To achieve convergence, PSO applies two types of knowledge, personal experience or *cognitive component*, which is the experience that every particle gets along optimization process; and social experience or *social component*, which is the experience that all the swarm get during the optimization process.

In simple terms, the particles are "flown" through a multidimensional search space where the position of each particle is adjusted according to its own experience and that of its neighbors [7].

Let $x_i(t)$ denote the position of a particle i in search space at time step t. The position of the particle is changed by adding a velocity $v_i(t)$, to the current position

$$x_i(t) = x_i(t) + v_i(t+1) \tag{1}$$

with. $x_i(0) \sim U(x_{min}, x_{max})$.

Originally two PSO algorithms have been developed which differ in the size of their neighborhood [7]. These two algorithms are called *lbest* and *gbest*. In the *lbest* algorithm the swarm is divided in neighborhoods of size n while in the *gbest* the neighborhood for each particle is the entire swarm, or in other words, the social component of the particle velocity update reflects information obtained from all the particles in the swarm [7]. For the *gbest* PSO, the velocity of particle i is calculated as

$$v_{ij}(t+1) = v_{ij}(t) + c_1 r_{1j}(t)\left[y_{ij}(t) - x_{ij}(t)\right] + c_2 r_{2j}(t)\left[\hat{y}_j(t) - x_{ij}(t)\right] \tag{2}$$

where $v_{ij}(t)$ is the velocity of particle i in dimension $j = 1,2,3,\ldots,n_x$ at time step t, $x_{ij}(t)$ is the position of particle i in dimension j at time step t, c_1 and c_2 are positive acceleration constants used to scale the contribution of the cognitive and social components respectively, $r_1(t)$ and $r_2(t) \sim U(0,1)$ are random values in the range [0, 1], sampled for uniform distribution. The random values introduce a stochastic element to the algorithm [7]. y_i is the best position that a particle i since the first time step and $\hat{y}$ is the best position found all the particles in the swarm.

Considering minimization problems, the personal best y_i is updated by equation (3)

$$y_i(t+1) = \begin{cases} y_i(t) & \text{if } f(x_i(t+1)) \geq y_i(t) \\ x_i(t+1) & \text{if } f(x_i(t+1)) < y_i(t) \end{cases} \tag{3}$$

where $f: \mathbb{R}^{n_x} \rightarrow \mathbb{R}$ is the fitness function.

The global best position, $\hat{y}(t)$, at time step t is defined by equation (4)

$$\hat{y}(t) \in \left\{ y_0(t), \ldots, y_{n_s}(t) \middle| f(\hat{y}(t)) = \min\left\{ f(y_0(t)), \ldots, f(y_{n_s}(t)) \right\} \right\} \tag{4}$$

where n_s is the size of the swarm. Table 1 shows the *gbest* PSO algorithm.

Table 1. *gbest* PSO

```
Create an initialize an n_x-dimensional swarm, S;
repeat
    for each particle I = 1,2,…,S.n_s do
        //set the personal best position If f(x_i(t+1)) < y_i(t) then
            S.y_i = S.x_i;
        end
        //set the global best position If S.y_i(t) < f(ŷ_i(t)) then
            S.ŷ_i = S.y_i;
        end
    end
    for each particle I = 1,2,…,S.n_s do
        update the velocity using equation (2)
        update the position using equation (1)
    end
until stopping condition is true;
```

3 Problem Statement

The model of the robot considered in this paper is that of a unicycle mobile robot (see Figure 1), that consists of two driving wheels mounted of the same axis and a front free wheel.

A unicycle mobile robot is an autonomous, wheeled vehicle capable of performing missions in fixed or uncertain environments. The robot body is symmetrical around the perpendicular axis and the center of mass is at the geometrical center of the body. It has two driving wheels that are fixed to the axis that passes through C and one passive wheel prevents the robot from tipping over as it moves on a plane. In what follows, it´s assumed that motion of the passive wheel can be ignored in the dynamics of the mobile robot presented by the following set of equations [8]:

$$M(q)\dot{\vartheta} + C(q,\dot{q})\vartheta + D\vartheta = \tau + F_{ext}(t) \tag{5}$$

$$\dot{q} = \underbrace{\begin{bmatrix} \cos\theta & 0 \\ \sin\theta & 0 \\ 0 & 1 \end{bmatrix}}_{J(q)} \underbrace{\begin{bmatrix} v \\ w \end{bmatrix}}_{\vartheta} \tag{6}$$

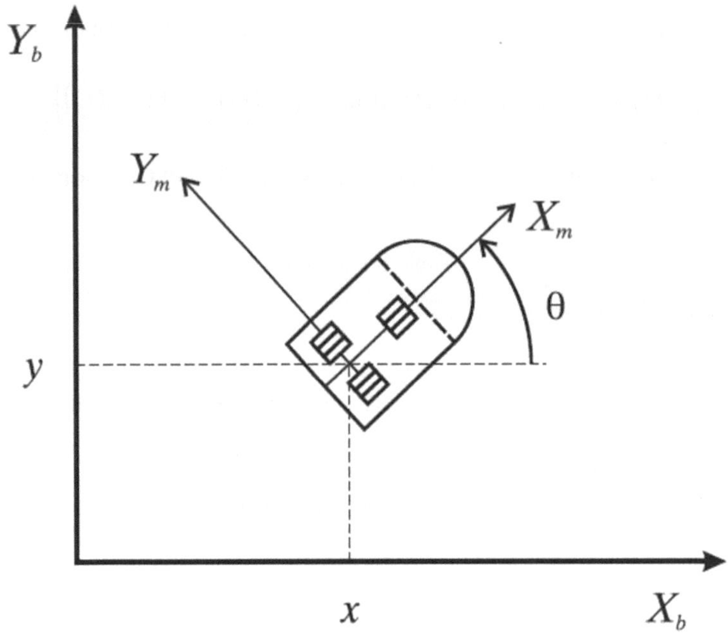

Fig. 1. Wheeled Mobile Robot [10]

Where $q=(x,y,\theta)^T$ is the vector of the configuration coordinates; $\vartheta=(v,w)^T$ is the vector of linear and angular velocities; $\tau = (\tau_1,\tau_2)$ is the vector of torques applied to the wheels of the robot where τ_1 and τ_2 denote the torques of the right and left wheel respectively (Figure 1); $F_{exr}\in\mathbb{R}^2$ uniformly bounded disturbance vector; $M(q)\in\mathbb{R}^{2x2}$ is the positive-definite inertia matrix; $C(q,\dot{q})\vartheta$ is the vector of centripetal and Coriolis forces; and $D\in\mathbb{R}$ is a diagonal positive-definite damping matrix. Equation (6) represents the kinematics of the system, where (x, y) is the position of the mobile robot in the X-Y (world) reference frame, θ is the angle between heading direction and the x-axis v and w are the angular and angular velocities, respectively.

Furthermore, the system (5)-(6) has the following non-holonomic constraint:

$$\dot{y}\cos\theta - \dot{x}\sin\theta = 0 \tag{7}$$

which corresponds to a no-slip wheel condition preventing the robot from moving sideways[9]. The system (6) fails to meet Brockett's necessary condition for feedback stabilization [2], which implies that non-continuous static state-feedback controller exists that stabilizes the close-loop system around the equilibrium point.

The control objective is to design a fuzzy logic controller of τ that ensures:

$$\lim_{t\to\infty}\left\|q_d(t)-q(t)\right\|=0 \tag{8}$$

for any continuously, differentiable, bounded desired trajectory $q_d \in \mathbb{R}^3$ while attenuating external disturbances.

A more detailed description can be found on reference [10].

4 Fuzzy Logic Control Design

In order to satisfy the control objective it is necessary to design a fuzzy logic controller for the real velocities of the mobile robot. To do that, a Takagi-Sugeno fuzzy logic controller was designed, using linguistic variables in the input and mathematical functions in the output. The error of the linear and angular velocities (v_d, w_d respectively), were taken as inputs variables, while the right (τ_1) and left (τ_2) torques as outputs. The membership functions used on the input are trapezoidal for the negative (*N*) and positive (*P*), and a triangular was used for the zero (*C*) linguistic terms. The interval used for this fuzzy controller is [-50 50] [10]. Figure 2 shows the input and output variables, and figure 3 shows the general FLC architecture.

The rule set of the FLC contains 9 rules, which governs the input-output relationship of the FLC and this adopts the Takagi-Sugeno style inference engine [10], and it is used with a single point in the outputs, this means that the outputs are constant values, obtained using weighted average defuzzification procedure. In Table 1 we present the rule set whose format is established as follows:

$$\text{Rule } i : \text{if } e_v \text{ is } G_1 \text{ and } e_w \text{ is } G_2 \text{ then } F \text{ is } G_3 \text{ and } N \text{ is } G_4$$

where $G_1..G_4$ are the fuzzy set associated to each variable i=1,2,…,9.

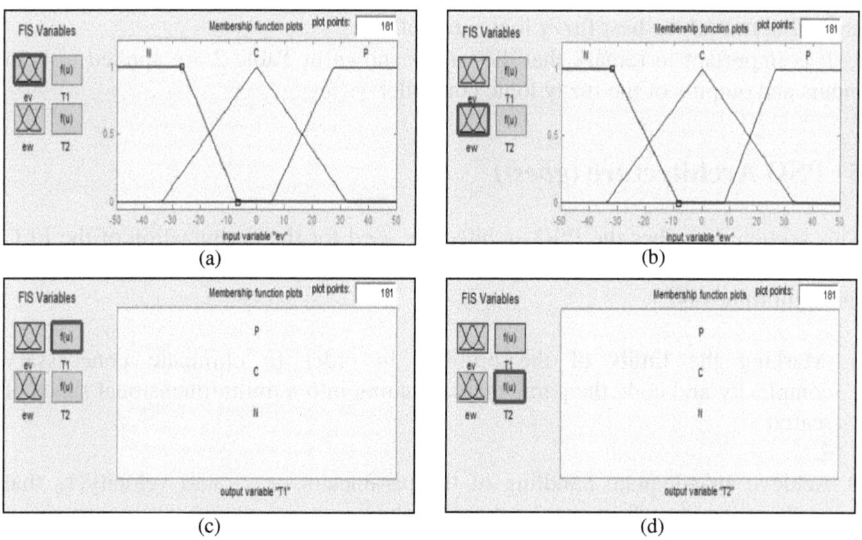

Fig. 2. (a) Linear velocity error. (b) Angular velocity error. (c) Right output (τ_1). (d) Left output (τ_2).

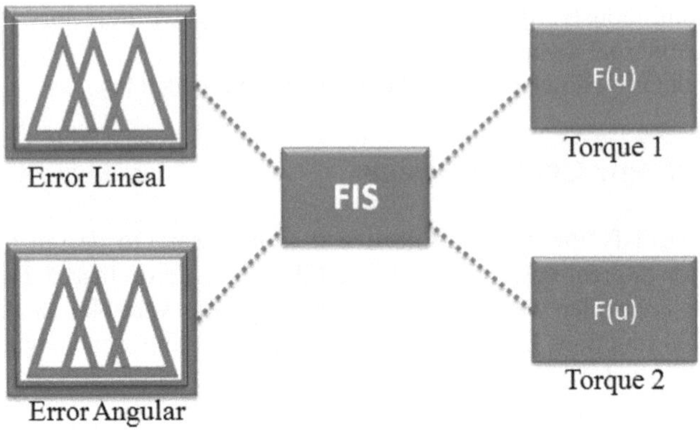

Fig. 3. Fuzzy Logic Controller Architecture.

Table 2. Fuzzy rules set

e_v / e_w	N	C	P
N	N / N	N / C	N / P
C	C / N	C / C	C / P
P	P / N	P / C	P / P

To find the best FLC, we used a *gbest* PSO to find the parameters of the membership functions. Table 2 shows the parameters of the membership functions, the minimal and maximum values in the search range for the *gbest* PSO algorithm to find the best fuzzy logic controller.

It is important to remark that the values shown in Table 2 are applied to both inputs and outputs of the fuzzy logic controller.

5 PSO Architecture (*gbest)*

This section describes the PSO architecture used for the optimization of the FLC for the autonomous mobile robot. The methodology applied for the developing of the algorithm was:

1. Marking the limits of the problem in order to eliminate unnecessary complexity and code the parameters to tuning into a multidimensional space of search
   2.
3. Achieve an adequate handling of the parameters c_1, c_2 and velocity v_{ij} that ensures a good performance of the algorithm.

5.1 Limiting the Problem and Coding the Problem

As we can see on Table 3, there exist several parameters that will not change at the end of the optimization process, for that reason we decided to ignore them and we reduced the number of elements that the method needed to find by deleting the elements whose minimal value and maximal values are the same. Table 4 shows the parameters of the membership functions included in the search.

Table 3. Parameters of the membership Functions.

MF TYPE	POINT	MINIMAL VALUE	MAXIMAL VALUE
Trapezoidal	a	-50	-50
	b	-50	-50
	c	-15	-5.1
	d	-1.5	-0.5
Triangular	a	-5	-1.8
	b	0	0
	c	1.8	5
Trapezoidal	a	0.5	1.5
	b	5.1	15
	c	50	50
	d	50	50
Constant (N)	a	-50	-50
Constant (C)	a	0	0
Constant (P)	a	50	50

Table 4. Parameters of the Membership Functions Included in PSO Search.

MF TYPE	POINT	MINIMAL VALUE	MAXIMAL VALUE
Trapezoidal	c	-15	-5.1
	d	-1.5	-0.5
Triangular	a	-5	-1.8
	c	1.8	5
Trapezoidal	a	0.5	1.5
	b	5.1	15

The next step was coding the parameters shown in Table 4 into a multi-dimensional space of search. Figure 4 shows the representation of a particle in PSO.

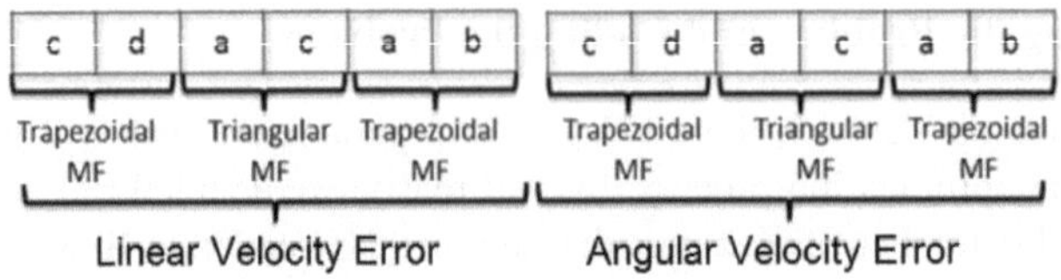

Fig. 4. Particle of the PSO.

5.2 PSO Parameters Handling

Before the application of the PSO, we needed to initialize the first swarm. Equation 9 shows the initializing method we used

$$x(0)=x_{\min,j}+r_j\left(x_{\max,j}-x_{\min,j}\right),\forall j=1,\dots,n_x,\forall i=1,\dots,n_s \tag{9}$$

where $x(0)$ is the matrix of dimensions, $x_{min,j}$ and $x_{max,j}$ are the minimal an maximum value respectively that a dimension j can take, and $r_j \sim U(0,1)$ is a random value.

Equation 10 shows the way we calculate the velocity and Equation 11 shows the maximum velocity

$$v_{ij}(t+1)=\begin{cases} v'_{ij}(t+1) & \text{if } v'_{ij}(t+1)<V_{\max,j} \\ V_{\max,j} & \text{if } v'_{ij}(t+1)\geq V_{\max,j} \end{cases} \tag{10}$$

$$V_{\max,j}=\delta\left(x_{\max,j}-x_{\min,j}\right) \quad \delta\in(0,1] \tag{11}$$

where $v'_{ij}(t+1)$ is the velocity calculated by equation (2), δ is a gain that is problem–dependent. For this parameter we used the heurist value of $\delta = 0.5$.

As we didn't know the adequate values for the constants c_1 and c_2 we decided to apply the method of Ratnaweera, who proposed that c_1 decreases linearly over time, while c_2 increases linearly [15]. This strategy focuses on exploration in the early stages of the optimization, while encourage convergence to a good optimum near the end to the optimization process by attracting particles more towards the neighborhood best (or global best) positions [7]. The values of $c_1(t)$ and $c_2(t)$ at time step t are calculated by Equation 12

$$\begin{aligned} c_1(t)&=\left(c_{1,\min}-c_{1,\max}\right)\frac{t}{n_t}+c_{1,\max} \\ c_2(t)&=\left(c_{2,\max}-c_{1,\min}\right)\frac{t}{n_t}+c_{1,\min} \end{aligned} \tag{12}$$

where $c_{1,max}$=$c_{2,max}$=2.5 and $c_{1,min}$=$c_{2,min}$ = 0.5.

Further improvements in performance have been obtained using a fuzzy self-adaptive acceleration scheme [16].

6 Simulation Results

In this section we present the results of the proposed controller to stabilize the unicycle mobile robot, defined by Equation (5) and Equation (6), where the matrix values

$$M(q)=\begin{bmatrix} 0.3749 & -0.0202 \\ -0.0202 & 0.3739 \end{bmatrix},$$

$$C(q,\dot{q})=\begin{bmatrix} 0 & 0.1350\dot{\theta} \\ -0.150\dot{\theta} & 0 \end{bmatrix},$$

$$\text{and } D=\begin{bmatrix} 10 & 0 \\ 0 & 10 \end{bmatrix}$$

were taken from [6]. The evaluation was made through a computer simulation performed in MATLAB® and SIMULINK®.

The desired trajectory is the following one:

$$\vartheta_d(t)=\begin{cases} v_d(t)=0.2(1-\exp(-t)) \\ w_d(t)=0.4\sin(0.5t) \end{cases} \tag{13}$$

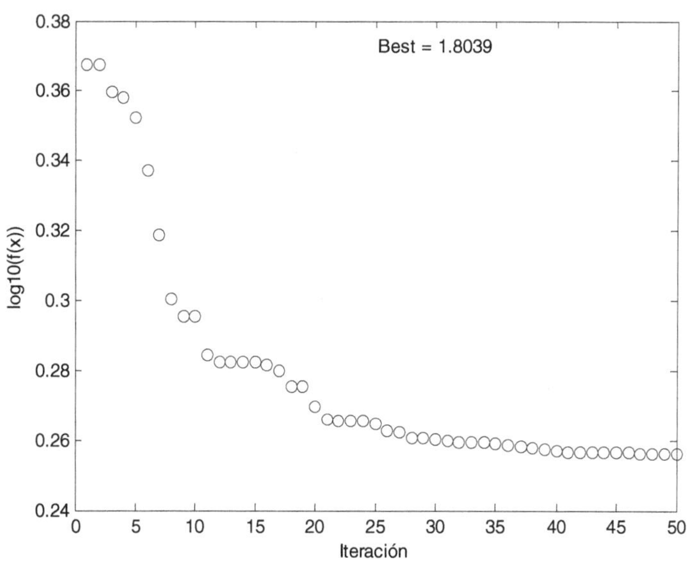

Fig. 5. Optimization Behavior of *gbest* PSO for Type-1 FLC.

and was chosen in terms of its corresponding desired linear v_d and angular w_d velocities, subject to the initial conditions

$$q(0) = (0.1, 0.1, 0)^T \text{ and } \vartheta(0) = 0 \in \mathbb{R}^2$$

The gains γ_i, i=1, 2, 3 of the kinematic model (see [10]) are $\gamma_1 = 5$, $\gamma_2 = 24$ and $\gamma_3 = 3$ were taken from [10, 17].

6.1 *gbest* PSO Algorithm Results for Type-1 FLC Optimization

Table 5 shows some results chosen randomly of the FLC, obtained varying the values of maximum iterations and number of particles, where the highlighted row shows the best result obtained with the method. Figure 5 shows the optimization behavior of the method for a type-1 FLC optimization.

Figure 6 shows the membership functions of the type-1 FLC obtained by the *gbest* PSO algorithm.

Table 5. *gbest* PSO Results of Simulations for Type-1 FLC Optimization.

Experiment	Iterations	Swarm	Average Error	Time
1	50	100	0.1523	00:09:08
2	15	150	0.1512	00:32:06
3	15	150	0.1516	00:33:03
4	15	25	0.1642	00:06:29
5	35	25	1.8981	00:04:18
6	**50**	**150**	**0.1509**	**01:00:50**
7	71	34	0.1515	00:02:35
8	71	34	0.1510	00:17:44
9	73	38	0.1520	00:03:19
10	250	50	0.1511	00:57:32
11	95	51	0.1912	01:16:20
12	51	57	0.1615	00:02:06
13	67	65	0.1728	01:29:28
14	67	65	0.1728	01:29:28
15	80	67	0.1994	01:46:27
16	65	68	0.1830	01:05:56
17	50	80	0.1591	00:18:24
18	78	82	0.1510	00:43:13
19	81	82	0.1510	00:40:09
20	17	91	0.1619	00:22:25
21	85	97	0.1510	01:31:51

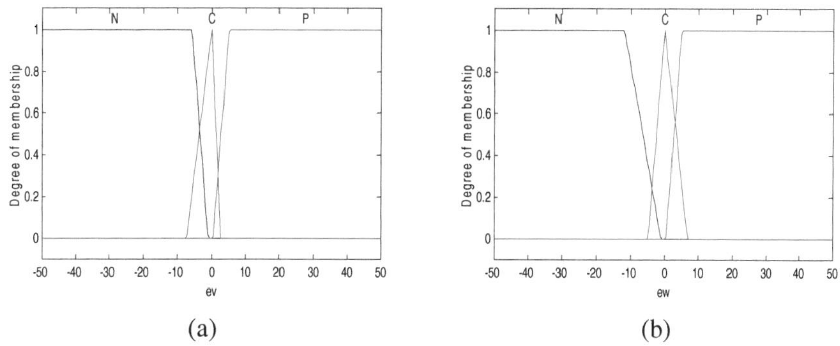

Fig. 6. (a) Linear velocity error, and (b) angular velocity error optimized by *gbest* PSO algorithm.

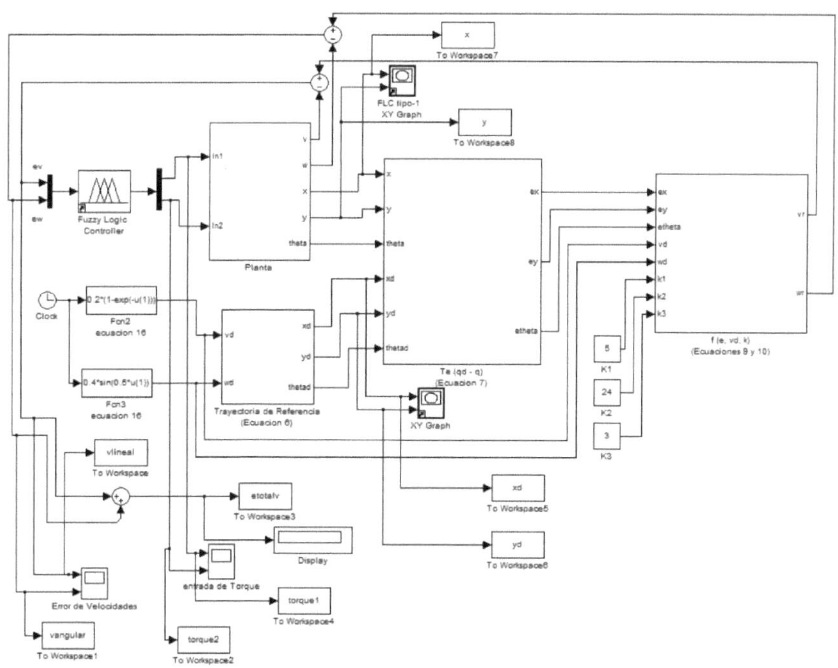

Fig. 7. Block diagram for simulation of the FLC

Figure 7 shows the block diagram used for the FLC that obtained the best results.

Figure 8 shows the desired trajectory and the obtained trajectory, where we can appreciate that they are very close.

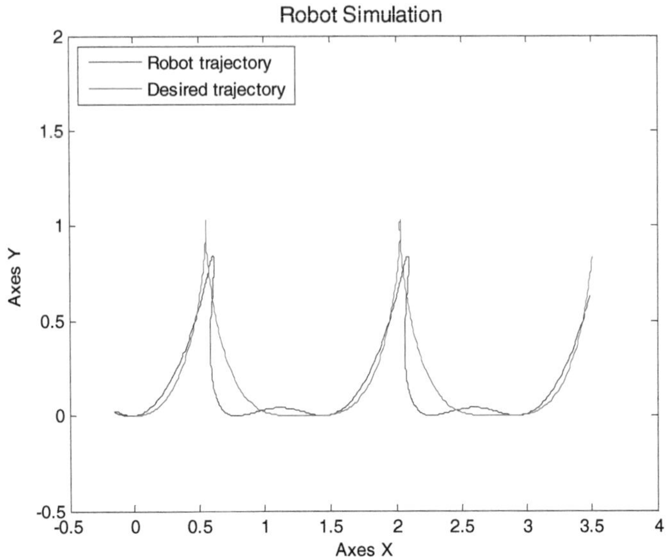

Fig. 8. Obtained trajectory with type-1 optimization

Table 6. *gbest* PSO Results of Simulations for Type-2 FLC Optimization.

Experiment	Iterations	Size of Swarm	Average Error	Time
1	10	10	0.1608	00:09:41
2	16	17	0.1607	00:16:32
3	21	17	0.1606	00:26:38
4	16	19	0.1614	00:18:06
5	28	20	0.1605	00:40:22
6	22	21	0.1607	00:25:25
7	28	22	0.1606	00:57:47
8	17	26	0.1606	00:24:22
9	19	26	0.1607	00:37:47
10	22	28	0.1606	00:42:26
11	27	28	0.1606	00:52:08
12	27	28	0.1606	00:51:56
13	28	28	0.1606	00:53:00
14	23	29	0.1608	00:39:09
15	71	34	0.1605	03:56:16
16	84	91	0.1601	07:46:58

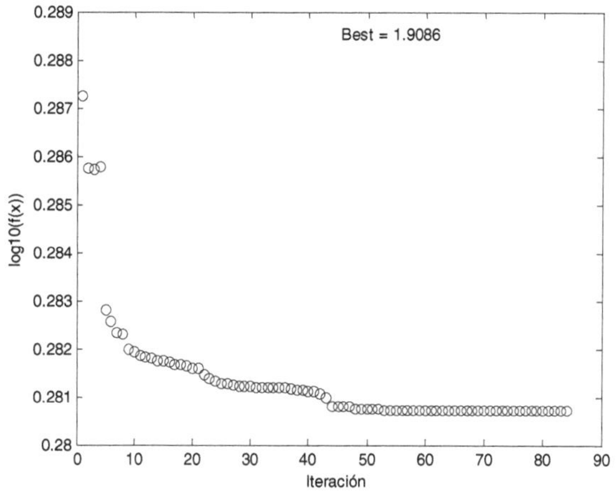

Fig. 9. Optimization Behavior of *gbest* PSO for Type-1 FLC.

(a)

(b)

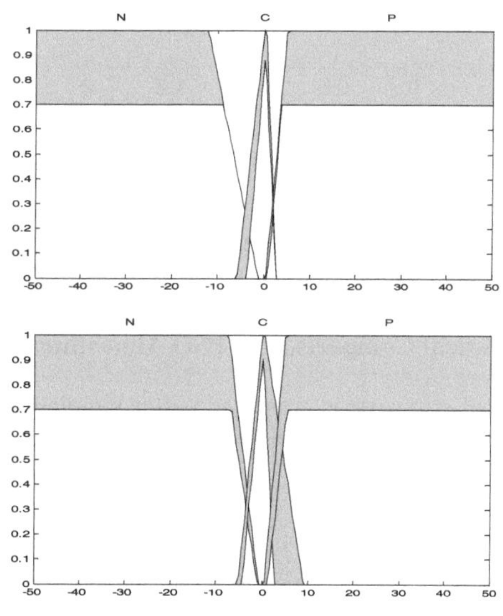

Fig. 10. (a) Linear velocity error, and (b) angular velocity error optimized by *gbest* PSO algorithm.

6.2 *gbest* PSO Algorithm Results for Type-2 FLC Optimization

Table 6 shows some results chosen randomly of the type-2 FLC, obtained varying the values of maximum iterations and number of particles, where the highlighted row shows the best result obtained with the method. Figure 9 shows the optimization behavior of the method for type-2 FLC optimization. Figure 10 shows the membership functions of the type-2 FLC obtained by the *gbest* PSO algorithm.

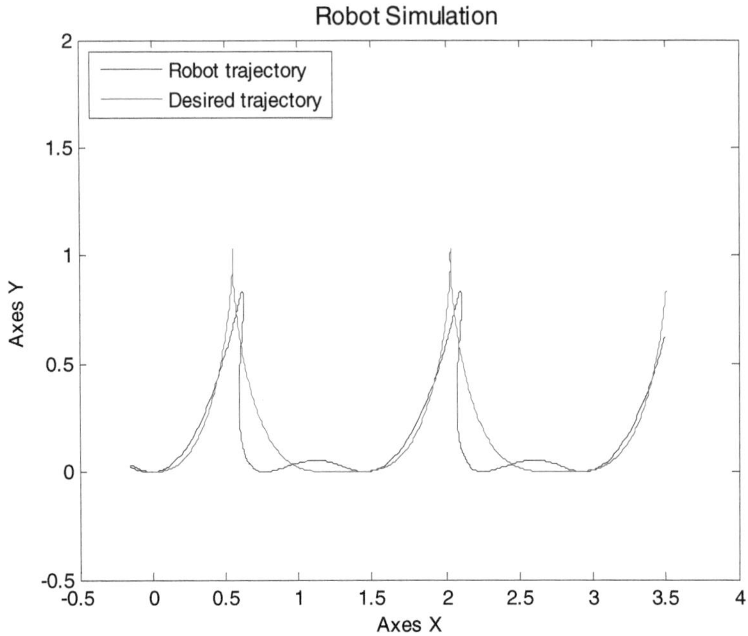

Fig. 11. Obtained trajectory with type-2 optimization

Figure 11 shows the desired trajectory and the obtained trajectory, where we can appreciate that they are very close.

6.3 Statistical Comparison of PSO Algorithms for Type-1 and Type-2 FLCs

A statistical comparison was made after simulations. To perform the comparison we applied hypothesis testing with t-student statistic. Equation (14) describes the t-student statistic sed.

$$T_0^* = \frac{\overline{X}_1 - \overline{X}_2}{\sqrt{\frac{S_1^2}{n_1} + \frac{S_2^2}{n_2}}} \tag{14}$$

where $\overline{X}_1$ and $\overline{X}_2$ are the means of the samples X_1 and X_2 respectively, S_1^2 and S_2^2 are the variance of sample 1 and sample 2 respectively and finally n_1 is the size of the sample 1 and n_2 is the size of the sample 2.

Table 7 shows the conditions of the hypothesis testing.

Figure 12 shows the value t = 5.0022 obtained by applying the hypothesis testing, where we can see that there is sufficient statistical evidence that allows us to say that PSO for type-2 optimization outperforms PSO for type-1 in the end of all experiments.

Table 7. Conditions for the Hypothesis Testing

Algorithm	Sample Size	Mean	Standard Deviation
PSO for type-1 optimization	175	0.3837177	0.5899010
PSO for type-2 optimization	49	0.1606571	0.0002179

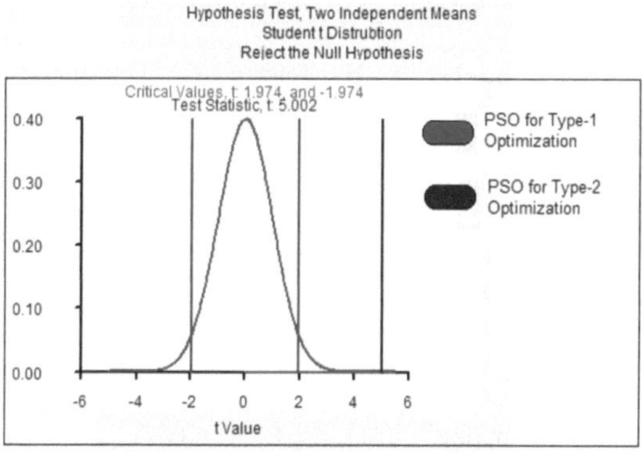

Fig. 12. Hypothesis Test

7 Conclusions

A trajectory tracking controller has been designed based on the dynamics and kinematics of the autonomous mobile robot through the application of PSO for the optimization of membership functions for the type-1 and type-2 FLCs with good results obtained after simulations. Finally we conclude that there exists sufficient statistical evidence to say PSO for type-2 optimization outperforms PSO for type-1 optimization although the best control achieved was obtained by one of the solutions with a type-1 FLC.

Acknowledgment. We would like to express our gratitude to the CONACYT and Tijuana Institute of Technology for the facilities and resources granted for the development of this research.

References

[1] Bloch, A.M.: Nonholonomic mechanics and control. Springer, New York (2003)

[2] Brockett, R.W.: Asymptotic stability and feedback stabilization. In: Millman, R.S., Susman, H.J. (eds.) Diferential Geometric Control Theory, p. 181. Birkhauser, Boston (1983)

[3] Castillo, O., Melin, P.: Soft Computing for Control of Non-Linear Dynamical Systems. Springer, Heidelberg (2001)

[4] Shi, Y., Eberhart, R.C.: Parameter selection in particle swarm optimization. In: Porto, V.W., Waagen, D. (eds.) EP 1998. LNCS, vol. 1447, pp. 591–600. Springer, Heidelberg (1998)
[5] Shi, Y., Eberhart, R.C.: A modified particle swarm optimizer. In: Proceedings of the IEEE International Conference on Evolutionary Computation, pp. 69–73. IEEE Press, Piscataway (1998)
[6] Duc Do, K., Zhong-Ping, J., Pan, J.: A global output-feedback controller for simultaneous tracking and stabilizations of unicycle-type mobile robots. IEEE Trans. Automat. Contr., col. 20(3), 589–594 (2004)
[7] Engelbrecht, A.P.: Fundamentals of Computational Swarm Intelligence, pp. 85–131. Wiley, England (2005)
[8] Lee, T.-C., Song, K.-T., Lee, C.-H., Teng, C.-C.: Tracking control of unicycle-modeled mobile robot using a saturation feedback controller. IEEE trans. Contr. Syst. Technol. 9(2), 305–318 (2001)
[9] Liberzon, D.: Switching in Systems and control. Birkhauser, Basel (2003)
[10] Martínez, R., Castillo, O., Aguilar, L.: Intelligent control for a perturbed autonomous wheeled mobile robot using type-2 fuzzy logic and genetic algorithms. JAMRIS 2(1), 1–11 (2008)
[11] Sepúlveda, R., Castillo, O., Melin, P., Montiel, O.: An Efficient Computational Method to Implement Type-2 Fuzzy Logic In control Applications. In: Analysis and Design of Intelligent Systems Using Soft Computing Techniques, June 2007. Advances in Soft Computing, vol. 41, pp. 45–52 (2007)
[12] Eberhart, R.C., Kennedy, J.: A new optimizer using particle swarm theory. In: Proceedings of the sixth international symposium on micro machine and human science, pp. 39–43. IEEE service center, Piscataway (1995)
[13] Kennedy, J., Eberhart, R.C.: Particle swarm optimization. In: Proc. IEEE International Conference on Neural Networks, vol. IV, pp. 1942–1948. IEEE service center, Piscataway (1995)
[14] Ratnaweera, A., Halgamuge, S.K., Watson, H.C.: Particles Swarm Optimizer with Time Varying Acceleration Coefficients. In: Proceedings of the International Conference on Soft Computing and Intelligent Systems, pp. 240–255 (2002)
[15] Ratnaweera, A., Halgamuge, S.K., Watson, H.C.: Particles Swarm Optimizer with Self-Adaptive Acceleration Coefficients. In: Proceedings of the First International Conference on Fuzzy Systems and Knowledge Discovery, pp. 264–268 (2003)
[16] Martínez, R., Castillo, O., Aguilar, L.: Optimization of interval type-2 fuzzy logic controllers for a perturbed autonomous wheeled mobile robot using genetic algorithms. Information Sciences 179, 2158–2174 (2009)

Author Index

MIX
Papier aus verantwortungsvollen Quellen
Paper from responsible sources
FSC® C105338

If you have any concerns about our products,
you can contact us on
ProductSafety@springernature.com

In case Publisher is established outside the EU,
the EU authorized representative is:
Springer Nature Customer Service Center GmbH
Europaplatz 3, 69115 Heidelberg, Germany

Printed by Libri Plureos GmbH
in Hamburg, Germany